Dennis Brouse on Horse Training

Bonding with Your Horse Through Gentle Leadership

★

By Dennis Brouse with Fran Lynghaug

Photography by Richard Hildreth

Voyageur Press

To my wife, Lori, my children, and my grandkids,
because of the wonderful love they give me, and to all horses.
—D. B.

First published in 2011 by Voyageur Press, an imprint of MBI Publishing Company,
400 First Avenue North, Suite 300, Minneapolis, MN 55401 USA

Voyageur Press titles are also available at discounts in bulk quantity for industrial or sales-promotional use. For details write to Special Sales Manager at MBI Publishing Company, 400 First Avenue North, Suite 300, Minneapolis, MN 55401 USA.

To find out more about our books, visit us online at www.voyageurpress.com.

Library of Congress Cataloging-in-Publication Data

Brouse, Dennis.
 Dennis Brouse on horse training : bonding with your horse through gentle leadership / by Dennis Brouse with Fran Lynghaug ; photography by Richard Hildreth. — 1st ed.
     p. cm.
 Includes index.
 ISBN 978-0-7603-4060-8 (flexibound)
 1. Horses—Training. I. Lynghaug, F. II. Hildreth, Richard, 1957- III. Title. IV. Title: Horse training.
 SF287B845 2011
 636.1'0835—dc22
                    2011006451

Edited by Danielle Ibister
Design Manager: LeAnn Kuhlmann
Designed by Cindy Laun
Layout by Erin Fahringer
Cover designed by Rick Korab, Korab Company Design

Printed in China

10 9 8 7 6 5 4 3 2 1

CONTENTS

MEET DENNIS BROUSE

DENNIS BROUSE'S HORSE-TRAINING SKILLS have been featured on television since 2002, first on the show *On the Trail Again* and later on his current program, *Saddle Up with Dennis Brouse*, which has become an exciting hit on national public television. As well as training lessons, the show features segments about interesting equine events, places, and breeds—from therapeutic riding to racing to draft horses. Dennis's newest enterprise is a six-part documentary series for public television called *The History of the Horse*.

Dennis Brouse truly knows horses. His approach to solving equine problems reaches the animal's mind and heart. He teaches how to create a connection with the horse, resulting in changes in its thinking, attitude, and behavior. His lessons make sense, are easy to follow, and—best of all—work on every type of horse and all kinds of problems.

What he teaches is fun, not work. Frustration dissolves as owners learn to relate to their horses and watch their animals transform before their eyes. People and horses come together with better communication and appreciation for each other, producing a strong partnership.

Dennis's methods are effective and innovative compared to other techniques. The best quality of his training is the extent of his gentleness with horses. The end result is a deeper understanding of the horse and the kind of carefree riding that is such a joy for any rider.

—*Fran Lynghaug*

Nick Vedros, MindFire Communications

THIS BOOK IS DIFFERENT than most instructional training guides, which typically are laid out in a precisely defined "you-should-do-this" procedure. Instead, Dennis Brouse describes each lesson with a horse as if you are standing there watching him. It's a simpler and easier way to understand how to train horses. He walks you through each session: what he does to correct a problem, how he does it, the idea behind it, and the results to expect. This method is designed to help you comprehend the lesson by showing what the exercise looks like and how long it takes. The idea is to teach your horse by doing what Dennis does and to learn during the process so that you achieve the same results.

Sprinkled throughout the lessons are his philosophies, insights, and explanations of his methods. You learn in the easiest way possible what to do with your horse for each problem and why the horse thinks and acts the way it does.

The book is divided into five main areas of basic horse training: (1) working in the roundpen, (2) controlling the horse for better riding, (3) solving common riding problems, (4) de-spooking for a safer horse, and (5) trailer loading. Each one contains its own methods and goals.

As a complement to the training information, the book includes intermittent features of interesting breeds and equine events, just like Dennis's television program. These features are designed to highlight the fun and joy of living with these magnificent animals and the reason why we want to train and connect with them.

—*Fran Lynghaug*

Nick Vedros, MindFire Communications

CHAPTER 1

THE ROUNDPEN

Hi! I'm Dennis Brouse. I appreciate having the opportunity to share my training philosophies with you. This book will show you lots of different horses with lots of different attitudes, from aggressive to pushy to disinterested. My goal is for you to have a happy and healthy relationship with your horse—a strong bond that lasts a lifetime. I want your horse to be a true partner, meaning it is both trusting and trustworthy.

The premise of my training is to get horses to do anything I ask as long as there's no hurt or pain involved. I don't "break" horses. I train them to be participating partners. Breaking horses with pain and intimidation will backfire eventually, and we don't want that. We want the end result to be a horse that understands, is calm and cool, has its eyes focused on us, and is responsive to our every movement.

INTRODUCTION TO THE ROUNDPEN

My roundpen technique is the hardest thing for me to teach of all my training philosophies. The typical roundpen method requires running the horse until it is lathered up. Any horse in the world will stop and look at you after too much of that, which is not the goal. I don't want to run a horse until it's sweaty because then the experience becomes negative. Establishing leadership and communication are the factors that form a true partnership, not exhaustion.

That is why roundpen work should be a positive experience for both horse and rider. Instead of running the horse a long time, I want to draw the horse in. This is reversed from everything we've been taught about the roundpen. When I use gentle methods to control speed and reverse direction, the horse recognizes me as the leader and wants to come to me. That is how communication and a friendly relationship begin.

Carefree riding is the end result of a horse that recognizes *you* as the leader.
Nick Vedros, MindFire Communications

I always introduce myself to a horse by extending a hand, which is the polite thing to do in a relationship.
Nick Vedros, MindFire Communications

Once leadership and control are established, the horse will either walk up to me when I draw it in or it will invite me to approach.

By being a kind and knowledgeable leader—meaning I'm careful and won't hurt either one of us—I gain the trust of the horse. This trust needs to be established before confronting any training problems, no matter what issues the horse may have.

We need to understand that horses are gregarious; for eons they lived in herds. Because of the way herds are structured, horses are always searching out a leader. Leadership is something every horse understands from the time it is a foal. When its mother looked at it, it knew exactly what she was communicating; it knew when she was scolding and when she was saying everything was okay.

Because of this gregarious nature, horses need to recognize us immediately as the leader or else they will assume leadership; for example, they will go where they want to go. They will lead us, and that's not good.

We need to establish ourselves as the leader in a way horses can understand—through basic communication, which should be a positive experience for the horse. This is the whole idea of the roundpen: using

We need to understand that horses are gregarious; for eons they lived in herds. Because of the way herds are structured, horses are always searching out a leader.

communication to establish leadership and a partnership with the horse.

The best way I have found to teach communication with the horse as well as leadership and partnership is by describing my demonstrations using different horses that have different problems. For that reason, I use examples of horses that have specific trouble areas to help explain my techniques. Let's get started.

ROUNDPEN OVERVIEW

The roundpen is the starting place of every horse I work with from two years of age and older. Even if the horse is well trained, I work it in the roundpen before I go riding. I can see better from the ground and more easily check the saddle and equipment or any lameness or health issues the horse may have before I get on its back. It also gets the horse's mind on me and focused on what is coming next.

The ideal roundpen is forty-five to fifty feet in diameter. If it is any smaller, I can't release the pressure; if it is any larger, it leads to exhaustion for both the horse and me.

Footing should be soft sand, not dirt, which can stir up too much dust. It should not be composed of hard surfaces, gravel, or grass, which would tempt the horse to stop and eat.

A roundpen should be forty-five to fifty feet in diameter. The round shape keeps horses from hiding in a corner.

> I want these equine partners of ours to make their own decisions in response to our leadership, but I also want them to discover that the best place in the world is beside us, which is something that wasn't there at the beginning of roundpen training.

Make sure the pen is round so that the horse can't hide in a corner, and make sure the fence is tall enough to prevent escape.

I control the horse's movement with a whip. It's not for punishment but for directional control only. A lot of people throw things at horses in the roundpen, like leadropes or lariats. I don't like throwing a lariat because I would lose control. Plus I would spend too much time gathering it back up. Also, a lariat would actually sensitize the horse, and I want to desensitize the horse.

THE LESSON

To start roundpen work, first the halter and leadrope are removed from the horse. Standing in the center of the pen, I make a kissing or clucking sound to get the horse moving; I might also have to lift the whip or swing it. Kissing or clucking along with gentle pressure from the whip will teach the horse to respond to the sound and move around the pen.

The horse will run away from me, usually at a trot, and circle the pen. I establish speed control by putting pressure on its hip. I face the hip and walk toward it, make a kissing sound, or use whatever it takes to get the horse to lope (**Figure 1**). You can use any cue you want. Your horse may need no more than a slight movement of your arms to go

into a lope, but other horses may really need you to step toward them or make a motion with the whip.

Once the horse begins to lope, I back off and assume a relaxed posture to slow it to the trot. If the horse doesn't slow down, I continue backing up and staying relaxed while calmly saying, "Relax." (That's why you need a big enough roundpen, so there's enough room to back off adequately.) Horses usually respond to this body language because they are constantly aware of our posture and what we are saying with it.

After allowing the horse to trot a little while, I get directional control by turning the horse around. I do this by moving back, which takes the pressure off the horse. I may need to move back as little as twelve inches, or I may need to move clear across the roundpen. I use this movement to ask the horse to come in toward me, which is the first step to turning it around. As it starts to turn toward me, I keep the horse moving so it will turn completely around and run in the opposite direction (**Figure 2**). I do this with a motion of either my hand or the whip.

Once the horse turns toward me consistently two or three times after circling a bit, I back off because I don't want to run it into the ground.

Good communication and leadership help you form a bond with your horse that lasts a lifetime.
Nick Vedros, MindFire Communications

When we can control the horse's speed and directional movement with these exercises of applying and releasing pressure, we gain the horse's confidence in us as the leader. Confidence is expressed when the horse carries its head low and licks its lips or chews and cocks an ear toward us while it runs, which is the beginning of it allowing us to be the leader.

After the horse signals its confidence in me, I withdraw all pressure by stepping back and saying, "Walk," which will draw the horse away from the fence and entice it to walk toward me (**Figure 3**). Some horses will stop and face me instead of coming to me, which is an invitation for me to approach them, and that is okay. Whether I approach the horse or it comes to me, I extend my hand slightly toward its nose to introduce myself.

Even if your horse is turning in toward you and following your verbal commands, you still want to make sure you are communicating properly. To verify that you are, don't make eye contact when you approach your horse or when it comes to you; instead, look at its knees

STEP-BY-STEP

Figure 1. The horse is going into a lope. You can tell she is not happy about it because her ears are back and she is wringing her tail.

Figure 3. After I've turned the horse back and forth a few times and her body language tells me she is confident in my leadership, I back off and say, "Walk," to invite her to come to me.

Figure 2. The horse has turned and is running in the other direction. You can tell she is more comfortable with me because now she has her inside ear cocked toward me.

Figure 4. The horse comes all the way to me. I introduce myself by extending my hand slightly, palm down, and not making eye contact.

and extend a hand toward its head. Each and every time you approach your horse, introduce yourself this way because it's the polite thing to do (**Figure 4**).

After introducing myself, I try to draw the horse in toward me more by using what I call "the invisible rope" (**Figure 5**). This is done by touching the horse on its cheek, closing my hand as if I have a leadrope in it, drawing my hand back, and stepping away from the horse. The horse should follow my hand as if it really has a leadrope pulling its head. Then I can turn and walk away and the horse will follow (**Figure 6**).

The invisible rope is mind control. When you can lead the horse around by the slightest movement of your hand, you have established your leadership (**Figure 7**).

ROUNDPEN SUMMARY

By gaining control like this on the ground in the roundpen using only the whip—not anything else—and getting directional control through pressure and release, you're allowing your horse to relate to you the first time you get on its back. The horse is going to understand the concepts: It will acknowledge you as leader and respond to pressure and release of pressure. Later, when you are on the horse and it does something you don't like, you can correct it by getting it to relax first and then using pressure to make the correction.

You can get your horse to relax with your body movement and tone of voice as you say, "Relax." By the time you are on its back, the horse will be familiar with this word and should relax.

I want these equine partners of ours to make their own decisions in response to our leadership, but I also want them to discover that the best place in the world is beside us, which is something that wasn't there at the beginning of roundpen training. The end result is our horses are comfortable being with us. That means they won't be hard to catch, regardless of all the distractions and other issues happening around them.

If you roundpen correctly and keep your patience, your horse will come around. So relax and enjoy it. Don't get frustrated. Draw your horse in, and always reward yourself and the horse by quitting on a positive note. With this philosophy, after a very short time you can do it and the results will be phenomenal.

Figure 5. Touching the cheek lightly with my fingertips and withdrawing my hand is using the "invisible rope."

Figure 6. I walk away with the horse following, which means leadership has been established.

Photos: Nick Vedros, MindFire Communications

Figure 7. I lead the horse away and it follows with its head right at my shoulder.

FOALS INSTINCTIVELY know from birth what their moms are communicating to them. Here, a Caspian colt and his dam enjoy a young visitor. The Caspian is an ancient breed from Iran that grows no larger than twelve hands. *Victoria Tollman, Equus Survival Trust*

THE HORSE THAT IGNORES

A horse that ignores the owner's commands is frustratingly hard to handle at best and downright dangerous at worst, especially when it comes to being around children. Ninety percent of the time, horse attitude problems originate from not doing proper work in the roundpen. So we're going to have fun in my favorite classroom, the roundpen, fixing a horse that has an issue with paying attention.

This particular mare couldn't wait to get away from her owner. When they first enter the roundpen, it looks like the horse is leading the owner more than the owner leading the horse.

All tack is removed. For any roundpen lesson to work, the horse needs to be free of tack and must feel free to move around the pen at will. The only way to establish leadership is to get control of the horse's movement when she has this freedom to move around.

Remember, we want a positive experience in the roundpen. We're not going to run the horse until she's lathered and tired enough to give in. Instead, we are going to form a partnership by first getting her moving and then controlling her movement. Ultimately, this will result in a much-needed attitude adjustment in the horse.

THE LESSON

The first thing I notice is the horse's attention on me is zero (**Figure 1**). When I kiss and raise the whip, however, she trots around correctly a couple of laps and drops her head a little bit once or twice, which displays a bit of submission. That's a decent place to start. What I am looking for eventually are strong signs that she is paying attention to me: licking, chewing, more dropping of the head, and her head is off the rail—that is, she is not turning her head toward the fence as she moves.

To keep the horse moving, I apply a little pressure by lifting and swinging the whip in a large circle at her hip to drive her on. The horse responds by shaking her head while she trots—a sign of aggression. She is saying, "I don't know what you want, but whatever it is, I don't like it!"

I drive the hip again, but she only shakes her head without increasing speed and pins her ears back, showing a bad attitude. If I push her too much now, she will only get worse. So I say, "Relax," lower the whip (**Figure 2**), and step away from her to turn her, which will help get control. She veers away from the fence and starts to turn around, facing me first; I drive at her hip to encourage the change in direction. She trots in the other direction and puts an ear on me (**Figure 3**). Good!

Next, I ask for the lope by making a kissing sound and raising the whip a bit. She ignores me at first, so I kiss again and swing the whip a bit more. She goes into a lope, but it is accompanied with head shaking. That's the same attitude my kids have when I tell them to clean their room, and it can happen with horses too. A horse's body language says a lot, and it's essential to understand it and have the right response.

What I really don't want to do is get aggressive with this horse and run her too much. I say, "Relax," and I back off to give her a chance to quit running and come in to me. I do this by raising my hand briefly in her direction, inviting her in. Instead, she makes a small turn (**Figure 4**) when she gets close and trots the other way.

STEP-BY-STEP

Figure 1. The mare is focused outside the pen and not paying attention to me.

Figure 2. I move away closer to the fence, to block the horse and turn her around. She veers away and turns by facing me first.

Figure 3. The mare trots a lap in the other direction and puts an ear on me.

Figure 4. When the mare gets close enough, she turns to go the other way, but she turns outward instead of inward.

Figure 5. I raise my hand toward the mare to invite her in to me.

The horse isn't doing what I want her to do. My intention is not to cut her off or drive her away. I want her to come in toward me, but she won't do it until she has to lope a bit. Changing speed when I ask for it will cause her to give in to me more, so I swing the whip to increase her speed back up to the lope. She shakes her head in disagreement again. There's an attitude problem every time I ask her to lope. If I keep the pressure up too long, I would really get the negative aspect of her attitude.

After I do a little more whip swinging and kissing, the horse does what I've been asking her to do: She goes into a lope. Her tail swishes, telling me she doesn't like it, but it's still a better attitude than before, so I immediately back away, tell her to "Relax," lower the whip, move closer to the fence, and kiss with my hand slightly raised toward her (**Figure 5**), inviting her to come in to me. I stretch my other hand out to block her path along the fence. Instead of coming toward me, however, she responds by turning around (although she faces me first, which is good) and trots off.

There is still an attitude adjustment to be made. The horse has her ear on me, but she's not coming in to me. I kiss and swing the whip, asking for the lope again, and she responds by kicking out toward me as she circles—a dangerous display of aggression.

I will try to fix the attitude issue that's happening every time I ask her to lope, but one time in the roundpen isn't going to fix everything. This training needs to be worked on consistently by the owner at home. There is no such thing as a quick fix. What we're going to try to do in this lesson is get a small attitude adjustment, a step in the right direction. If she kicks out again, I will correct her with the whip and make her run. It's like when I played football in college: If I did

Attitude problems disappear when a horse is able to relax and when you can control its movement through changes in speed.

something wrong, I had to run laps, which is how I learned my lessons.

When I swing the whip again and kiss to her, asking her to lope, she responds by wringing her tail; she's telling me she's uncomfortable and doesn't want to relinquish leadership. She tosses her head, another little bit of an attitude, and so I wait for that to dissipate. As soon as it does, I'm going to back off to reward her for calming down. If I kept the pressure up, her bad attitude would continue or possibly get worse.

The next time I ask her, she turns around with little trouble. Her ear is on me, which is a good sign, but I would like to see her chewing. I'm not going to do anything more until she asks me what I want. Instead, she does a little tail swishing, which is still a dominant attitude; she doesn't want to pay attention. She's telling me, "I want nothing to do with you. I'm the leader!"

She kicks out at me a second time. I respond by making a kissing sound and "spanking" her just a little, which means I swish the whip toward her heels. I reserve spanking only for a horse that does something dangerous like kicking or threatening to kick. It won't hurt the horse. It's imperative that the force used with a whip doesn't increase in severity according to an increase in the horse's bad behavior. The tactic changes, not the energy to enforce it.

The horse responds by loping without the attitude she had previously (**Figure 6**) and even dips her head down in submission. Good! Because she responded to my command and I don't want her out of breath or sweating, I say, "Relax," and raise a hand in her direction. She stops and faces me, and I change the whip to the other hand and drive at her hip to complete the turn. She does a nice, even turnaround and trots off in the other direction.

Again I ask for the lope, and again she wrings her tail. That is the same display of a bad attitude, so I spank her again. I corrected her because if it escalates and she kicks out or bucks again, it's dangerous.

After she lopes, I tell her "Relax" until she slows and faces me as she makes her turn-around, and I only raise the whip enough (**Figure 7**) to encourage her on in the other direction.

Attitude problems disappear when a horse is able to relax and when you can control its movement through changes in speed. In order to make this a positive experience, I'm going to see if I can control the horse's movement without the whip, which is less pressure, and start getting her to truly relax. That way she'll understand the leadership issue and back off of it. I kiss to her, lower the whip, and back away. She slows and watches me, so I direct her to turn around slowly and go the other way with a slightly raised hand, but she walks off too quickly after her turnaround.

She is still nervous, so I do it again: I raise a hand, inviting her in, and then point in the other direction. She stops, faces me, than walks off in the other direction. Perfect! It was a more gentle transition for her, which is what I am looking for. She is now chewing and dipping her head toward the ground, showing submission. I direct her to turn with only a slight movement of my hand, not the whip. She is watching and turns slowly, which is very good.

It's time to ask her to trot in response to my hand movement alone, no whip. I kiss to her and only use a slight motion of my free hand. She responds well and trots off. Nice! I reward her by dropping the whip altogether. I might have quit too quickly, but we'll find out.

Now it's time to try to draw her in. I say, "Relax," and raise my hand with arm bent and fingers pointed down. She stops trotting, faces me, and waits (**Figure 8**). I step closer with my shoulder in toward her and then stop to wait for her to invite me in. She turns her head away, so I continue to wait. Then she turns her head back and faces me, which is saying that it's okay to come closer. I approach her again and raise my hand casually toward her. When I get close, she turns her head away. I stop and draw back my hand while backing away, which releases the pressure for her to accept me. She faces me, inviting me in again, as long as I am not pressuring her. Cool!

I am able to approach and let her sniff my hand, palm down (**Figure 9**), but she keeps turning her head away from me. Every time she does, I kiss to get her attention and back away. When she faces me, I approach. With this back-and-forth approach, I am able to extend my other hand, and she sniffs that one too.

She is still a little nervous and fluctuates between ignoring me and looking at me. She hasn't quite decided what to do. I touch her lightly on the neck, kiss, and back away at an angle to her rear with my hand raised by her head, inviting her in. She turns and faces me. Perfect! But then she puts her head down, ignoring me, not wanting to relinquish her leadership just yet.

STEP-BY-STEP

Figure 6. The mare lopes off with a better attitude.

Figure 7. I raise the whip only enough to encourage the horse to complete the turn and move in the other direction.

Figure 8. I raise my hand to the mare with my arm bent, fingers pointed down, and she faces me.

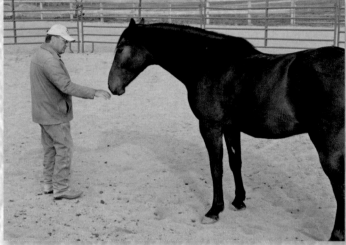

Figure 9. I approach when the mare is facing me and introduce myself. She isn't sure if she likes this.

Figure 10. The mare ignores me and nibbles on the ground.

She wants to disregard me and I won't tolerate that, but I won't punish her. I just back off and let her try again. (**Figure 10**). If she walks away, then I'm going to make her run. She will think, "What did I do wrong?" and remember that the last thing she did was walk away.

We are close to establishing who the leader is. We go back and forth with little tests between her ignoring me and my motioning to drive her off:

- I can stroke her neck a little, but she puts her head down, intentionally ignoring me.

- I kiss to her to bring her head up and then back away with my hand raised, drawing her in. She ignores me with her head down, nibbling on the ground.

- I stomp my foot at her to get her head up and approach her again. She raises her head and I stroke her neck, but she turns away.

- I kiss and back away again with a raised hand. She nibbles on the ground and ignores me.

- I stomp my foot and raise my hand to shoo her away and get her attention. She continues to ignore me.

- I step back and wave my cap at her to drive her off (**Figure 11**). She decides she doesn't really want to run, so she turns to face me.

Because she's free of any tack and can move at will, it's her choice if she wants to leave me, ignore me, or face me—but if she leaves or ignores me, I will drive her off and make her run until she decides to face me. This is the best way to handle the attitude issue.

After these back-and-forth maneuvers, she starts responding to me better. Now I can approach her head on. She is paying attention to me and doesn't turn away. I avoid eye contact as I approach her, which can drive off some horses.

I make an attempt at the invisible rope (see Overview, page 16), stepping to her side and drawing the rope away from her head so she will follow me. Instead, she ignores me, so I take off my hat and shoo her; it's her choice to do what she wants. She meanders a bit, not really paying attention. I swish my cap hard at her to drive her off, and she decides again to face me. I approach from her side with a raised hand (**Figure 12**), then back off, drawing her to me. She takes a step toward me, so I approach and touch her lightly on the cheek, then walk past her. She follows for only a few steps, then ignores me by putting her head down, nibbling on the ground, and I stop right away. When I back up at her side, she backs up with me, but when I walk forward again, she ignores me. I want her to respectfully follow me without out use of a leadrope. If a horse follows the trainer without being led, that means leadership has been established, which is the whole purpose of roundpen work. But I don't get that, so I drive her off until she pays attention and starts respecting me more.

This back-and-forth testing between us continues for a while: I invite her in, and she only stays briefly, then ignores me. When I motion to chase her off—stomping a foot and waving my cap—she returns to watching me. Finally she points her ears at me and squares herself up next to me. There! She's finally paying real attention.

The more we do this, the better she'll get, but it will take a lot of additional sessions for her to become reliable. This groundwork is so imperative because of the attitude she had ("I don't want to have anything to do with you, so I'm going to put my nose on the ground and ignore you"), but I won't tolerate that. Allowing this attitude would only support her to continue it, and I would be turning over leadership to her again.

I kiss to her and stroke her neck to get her attention (**Figure 13**), and she turns her head toward me, which is what I am asking, but she doesn't hold it long. She knows what I want; she just doesn't want to relinquish it quickly, which is okay for now. I step to her side and use the invisible leadrope to draw her in to me, and this time she follows me (**Figure 14**). Cool! Her head is low when I lead her and her eyes are calmer, which means she is relaxed because she is no longer in charge.

SUMMARY

I didn't use heavy punishment in this lesson and I still got the horse's respect and submission. The worst thing I did was spank the whip at her heels, which didn't hurt her. The roundpen lesson has to be a positive experience for the horse or we won't get the right response. Her attitude could have escalated into something worse if I was too hard on her.

This is a difficult thing to teach: when to allow a horse to express its attitude and when to correct it. Expect gradual changes after a few laps in the roundpen. Read what the horse is saying by its cues. Signs of submission are when a horse turns its ear on you, licks, chews, dips its head down, and turns around by facing you first.

Establishing leadership is essential to dealing with a horse that ignores the owner and lays the foundation for any further training. Most horses don't want to be leaders, especially in the beginning. They are always looking for a leader and will only assume the position if you don't establish yourself as a caring and knowledgeable leader. It's important to show that you understand their body language and, though you demand

> **It's important to show that you understand their body language and, though you demand their respect, you can be trusted not to hurt them. That's being a leader.**

their respect, you can be trusted not to hurt them. That's being a leader. Horses like that kind of authority and respond positively to it; it's natural for them. They don't have to be afraid or suspicious of danger. They relax and turn over control once a pecking order is in place.

Now that the horse has responded well to me, this is a good time to put a halter and leadrope on her, invite the owner in, and have her do some of these methods with her horse. I coach her on how to do them.

Some ground rules I share with her are: Each time you approach your horse, do it exactly the same way, whether you are going to feed her or ride her or whatever. Don't walk up with what the horse will interpret as an aggressive attitude: shoulders squared and approaching quickly, boldly, or directly with arms swinging. That would scare anyone. Approach quietly. Let the horse smell your hand (**Figure 15**), then back away toward her hip, inviting her to turn and face you. If she doesn't, or if she turns away, drive her off. If she faces you, it's okay to approach.

When you are haltering a horse and it has given the signal to approach, extend your hand, tell the horse "Relax," and put the halter and leadrope on. If your horse really respects you as a leader, she will respect your

Figure 11. I wave my cap at the mare to drive her off, but she doesn't really want to run anymore.

Figure 12. I approach the mare with a raised hand and she pays attention.

Figure 13. I stroke the mare's neck to get her attention.

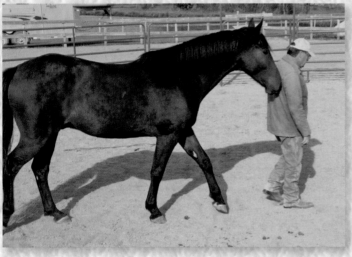

Figure 14. The horse is finally following me without any leadrope. Leadership is established.

Figure 15. The owner introduces herself and approaches the mare with a halter and leadrope. The mare faces the owner and doesn't turn away.

space, such as not walking across your personal area and staying a step away.

When leading your horse, be consistent with whatever distance you want her to be from you. If she gets too far ahead, back up and kiss with a loose leadrope, inviting her in to you; don't tug on her. Have her square up facing you. Walk off again with a loose lead and do a tight turn to see if she gets in your space; correct her if she does. If she avoids getting too close to your shoulder and moves her head around it, she is respecting your space.

How you hold the leadrope is important. Don't hold it close to the halter. Hold it loosely and walk off confidently. If you hold it too tightly, it's transmitting to her that you are nervous. Walk off and don't pay too much attention to the horse. Look straight ahead, trust that she's there with you, and stop when you want to. If the horse lifts her head and looks outside the roundpen, she's not paying attention, so before you ask her to follow you, get her attention back. You will know you have her attention if her ears are on you. Then cluck or kiss to her and turn in a circle.

We'll establish leadership the more we work in the roundpen. It won't happen with one lesson, but we've got a good start. The more time we spend on groundwork, the less time we'll have to spend on the ground; we don't want to get bucked off!

The horse's eyes are completely changed; she has a lot more respect for the owner. She's paying attention and respecting the owner's space, so it's a good time to quit and let her relax. The last thing she's going to remember about being in the roundpen is walking peacefully at the owner's shoulder.

| Total time for this session: 16 minutes |

Dennis directed these exercises:
1. Trotted the horse in each direction.
2. Loped the horse 4 times.
3. Turned the horse 10 times.
4. Corrected the horse from wringing her tail and kicking out.
5. Got the horse to focus on him and square up her stance next to him.
6. Corrected the horse from turning away and ignoring him.
7. Had the horse follow him using only the invisible leadrope.
8. Gave a lesson to the owner about keeping the horse's attention while leading it on a loose leadrope.

BONDING IS an important part of trust that aids in leadership and maintaining control. This girl enjoys a special moment with her family's Fell Pony stallion, a rare breed from northern England's Lake District. *Dwain Snyder*

STEP-BY-STEP

Figure 1. The horse moves closer to his buddy, who is tied a distance away, and ignores me.

Figure 2. The whip is lowered, but the horse turns by facing the fence first.

Figure 3. The horse runs with his head angled out toward the fence, looking for his buddy.

Figure 4. I cut the horse off, trying to turn him the right way around (by facing me first), but he turns outward instead.

THE HORSE THAT IS BUDDY SOUR

When a horse is more focused on another horse than listening to you, it is "buddy sour." The horse wants to be with its buddy, not you. It doesn't yet trust you or recognize you as the leader.

In the roundpen, when asked to change directions, a buddy-sour horse will turn around by facing the fence first to scan for its buddy if the buddy is outside the roundpen instead of turning inward to face you first, as the horse is supposed to do. The horse will also move around the roundpen with its body language attuned to the buddy: body angled toward the fence, head pointed outside the roundpen, and an ear on the buddy instead of you.

My goal in this lesson is to get the horse's concentration off its buddy and onto me, the leader. As always, I like to work on control techniques in my "classroom study," the roundpen. This is where I establish leadership with problem horses by controlling their movement, which is the foundation of communication as well as leadership.

Today I have an owner bringing her gelding into the roundpen for me to work with while his buddy, a mare, is standing tied outside the pen some distance away. I ask the owner to walk her horse around so I can watch him first. Then his halter and leadrope are removed and she leaves.

THE LESSON

I begin by raising my hand to start the gelding moving and right away he shows separation anxiety. He leaves the center of the ring immediately, walking to the side of the pen to be closer to the other horse outside the pen (**Figure 1**).

I want the horse to relax and focus on me as soon as possible and not the other horse, so I ask him to circle the pen at a trot by swinging my whip, and he trots off. My goal is to control his movement and look for specific cues from him: ear cocked toward me, licking, chewing, softening of the eyes, and eventually always turning in toward me when reversing directions after I ask for it. When a horse gives me that, it means, "What do you want?" It means he's paying attention to me, which is what I am trying to get.

The gelding makes a couple of laps around the pen before I ask him to lope by swinging the whip more and making a kissing sound to him. He responds well and goes into a lope so I immediately back off, lower the whip, bring my hands down, kiss to him, and raise one hand toward him. This is my way of inviting him to turn in toward me, but he turns by facing outward at the fence (**Figure 2**) and lopes in the other direction. That's okay for now; I am controlling his speed and he turned right away, which are good things to build on.

His ears are on me, so I back off and kiss to him. He stops and turns around again by facing the fence first. As he trots off, I notice that he's not moving parallel to the fence. Instead, his head and body are angled away from me and toward the fence where he can look for his buddy (**Figure 3**). Having his head "on the rail" like this means he does not yet acknowledge me as the leader.

This horse is doing a few things right (trotting and loping when I ask him to, keeping an ear on me) and a few things wrong (keeping his head on the rail and turning outward when he changes directions). I ask for the lope again by swinging the whip more and kiss to him, and he easily goes into a lope. I back up and lower the whip, inviting him to face me. Instead, he turns again by facing the fence first; it's the wrong way, so I stop him and turn him again by stepping toward him. Again he turns the wrong way and lopes off. I say, "Relax," and withdraw pressure by stepping back, inviting him in to me, but he continues and runs around behind me—between the fence and me. I am able to turn him again several times, but each time he turns the wrong way.

This means that the horse has a little bit of an attitude that needs adjustment. He's not being mean, he just isn't recognizing me as the leader. I need to be consistent with him and help him realize that he can trust me and follow my lead. I tell him to relax and then I back off, kiss to him, and raise a hand toward him. For the first time, he turns by facing me first, which is what I want, and goes the other way correctly. Cool!

Now his ears are on me, he is chewing and actually submitting, but he's still angled toward the fence a bit too much.

Here I make a little mistake. When the horse shows these signs of submission, I am not quick enough in rewarding him by slowing him down and inviting him to come in toward me. So when I try to turn him again, it's too late and he turns the wrong way. That is my fault. It's hard sometimes to respond quickly enough when the horse is running around the pen and starts displaying cues of submission; it can be missed easily. We all make mistakes, and we can't always blame the horse. You have to really look for the cues, such as chewing and ears turned toward you, to make the right response at the right time. Some people totally miss the cues and make their horse keep running.

It's important to be persistent at times like this when the horse can turn correctly but is insisting on doing it wrong.

I say, "Relax," and back away while lowering my hands and blocking his way by the fence. He turns the wrong way again, but I want to give him another chance, so I move across the ring and cut him off there to turn him (**Figure 4**). Nevertheless, he turns the wrong way again.

It's important to keep giving the horse a chance to do the right thing. He's chewing and acting submissive, so I kiss and back away, raising a hand toward him, and this time he responds by turning the right way, facing me. I help him go the other direction by pointing with my free hand, and he completes the turn and goes in that direction. There! Very cool!

The horse is giving me more and more submissive cues, except he is still angled toward the fence as he moves around. I tell him to relax and then I step in toward the fence, blocking his circling. When he starts to turn the wrong way, I switch tactics and step toward his rear (**Figure 5**) to prevent it from swinging out toward me, and he continues trotting the same direction without making the turn.

I want to try to intercept any incorrect turns like this instead of allowing him to complete them. I need to catch him before we both have to do more work.

STEP-BY-STEP

Figure 5. I step toward the horse's rear to prevent him from turning the wrong way, but he only continues trotting without turning around.

Figure 6. The horse slows down and looks at his buddy when he gets close to her.

Figure 7. The horse keeps stopping by his buddy and turning the wrong way.

Figure 8. The horse is starting to turn in toward me to do the correct turn-around.

Figure 9. The horse has submitted somewhat but is still a bit apprehensive.

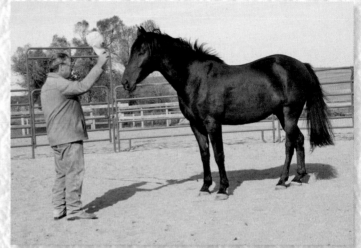

Figure 10. I wave my cap at the horse, but he continues to stand there.

Figure 11. The horse chooses to come to me, even with his buddy close by.

I drive him on and then ask him to turn in toward me again by backing off and raising a hand toward him, but he wants to be with his buddy so much that he turns outward when he's closest to her instead of inward toward me when I ask him. This is actually good because the more distractions there are, the better off we will be if I can draw him away from where she is standing.

I'm going to try and control him by turning him correctly in that exact spot where his friend is. I kiss and back away, inviting him in, but he puts his head over the fence (**Figure 6**), trying to be closer to the other horse every time he circles close to that area. This horse is a classic example of buddy sour.

I say, "Relax," and withdraw pressure, but he turns the wrong way again. I stop him from continuing in that direction and turn him around. I try several times to turn him correctly and he keeps stopping by the other horse (**Figure 7**). Every time he turns the wrong way, I stop him to correct him. When I pressure him to keep loping, he kicks a little in my direction, which expresses what he thinks.

This behavior is not aggressive; he is just getting frustrated. I continue asking him and he finally turns around correctly by facing me first right at the same spot where his buddy is, and then he moves off in the other direction. Cool! If I can get him to turn correctly one more time, maybe he'll submit to me.

I kiss, lower my hands, step back and away from him, raise one hand slightly in his direction, and he turns around correctly by facing me first, but does it with another little kick. I ask him again and this time he does a nice turn the right way.

The reason why I had him turn again is because he didn't show enough submission the

first time. He kicked out, which means he was reluctant to turn that way. I want a nice, calm, cool turn in toward me. I want him to say, "Dennis, I'm all yours. What do you want?"

I tell him to relax, kiss to him, and he turns again, but it's the wrong way, toward the fence. I rush to stop him and turn him back the other way, really making him lope now. Then I kiss, back away, and raise my hand toward him, but he turns the wrong way again, this time with ears back and tail swishing. Each time I stop him and turn him around, I make him lope. We do this several times and each time he turns the wrong way, but I want to keep giving him a chance. When I swish the whip to make him lope, he kicks out again, which is one good reason why it's not good to have the horse's rear facing us when we turn him.

It's important to be persistent at times like this when the horse can turn correctly but is insisting on doing it wrong. It only takes one more time for me to intercept him as he turns incorrectly before he does a nice turn in toward me and goes the other direction (**Figure 8**). Great!

My intent is not to work him until he gets sweaty and tired, it's just to gain mental control. I kiss to him, step back, and raise a hand to invite him to come to me, and he turns in toward me quite well and walks up to me.

The horse is submitting to me now, so I drop the whip. I lose eye contact to keep him calm and comfortable. He is standing close to me (**Figure 9**) and chewing with his eyes looking much calmer. I extend both hands and he sniffs briefly. Then he turns his head away, diverting his focus to his buddy, but when I kiss, he turns back to me. I have regained his attention, even with his buddy standing there.

I step back with a hand raised, kissing and inviting him to follow, but he looks outside the pen. He's uncertain. If he decides to walk off, I will drive him away.

I step back again with my hand raised, inviting him in, but this time he steps past me. I respond by stomping my foot and swishing my cap at his hip. This sends a clear message that he made the wrong choice. He walks off, facing the fence, and stops. He is distracted and indecisive. It was his choice to go where he wanted, but it's my job to pressure him to make the next choice the right one.

I want horses to make their own decisions, but I will help them make the right one that works for both of us.

I want to give him another chance, so I approach with my hand raised, kissing to him and inviting him in again. He only acknowledges me a little. I try to drive him off by waving my cap (**Figure 10**), but he really doesn't want to run again and just stands there. I'm not connecting with him just yet.

This horse needs a little bit more of an attitude adjustment, so I go back for the whip and use it to make him run again, clicking to him to move out. He goes into a trot, circling the pen, and I raise my hand to him, drop the whip, and step back after only half a lap. He slows and faces me (**Figure 11**), again walking up to me while chewing and licking—signs of submission. It's his choice if he wants to come to me or be driven off, and he chooses to come to me.

I introduce myself again, hands raised palm down, which is something that should be done every time we get close to a horse. Be consistent with this. It's polite and only takes a few seconds. We should always introduce everything to the horse—ourselves, the saddle, the bridle, or any other equipment.

Now that the gelding is being more submissive, I am intentionally going to try to draw him away from his buddy who is standing nearby. I tell the horse to relax, and while he stands quietly, I stroke his head and neck. I step to his side and kiss, then back away at an angle to his hip with my hand raised. He swings his hip away and turns to me, following my hand (**Figure 12**). That is so cool! I asked him to follow me away from his buddy and he did.

I kiss and walk away and he follows. I stop and tell him to relax. This is another chance for him to look at his buddy. It's okay if he looks at her as long as I can regain his attention.

He looks at the other horse and takes a step forward. If he walks away, I'll have to correct him. I step toward him and put my hand on his outside cheek to direct his head toward me. Then I put a hand close to his eye and draw it back toward me (**Figure 13**), which is using the invisible rope (see Overview, page 16).

He puts his head down, ignoring me. This was the wrong thing for him to do, so I motion to drive him off by waving my hands and then my cap at him, but he refuses to move off. I don't have him mentally on this side of his head yet. Some horses will connect to me from one side of their brain but not the other. If he continues to ignore me, I'm going to run him off.

I try to use the invisible rope to draw him away from his buddy again, but she's pulling him as hard as I am. I want to make sure that I can pull harder, so I kiss, stroke his neck, and raise my hand with the invisible rope, but he ignores me. (**Figure 14**).

He is still not connecting completely to me, so one more time, I pick up the whip and drive him off (**Figure 15**). He gives a little kick

STEP-BY-STEP

Figure 12. I raise my hand and the horse responds by moving his hip away from me.

Figure 13. I touch the horse's jowl with my fingertips near his eye. This is using the "invisible rope."

Figure 14. I draw my hand back toward me after touching his jowl.

Figure 15. I drive the horse off because he is too focused on his buddy.

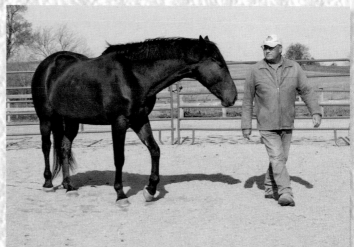

Figure 16. The horse walks along with me without a leadrope.

Figure 17. When I direct a hand at the horse's hip, I can make it move away and turn its head toward me.

on the off side in objection. He just told me what he thought again, but I don't want to run him too much. I like to give horses a chance to make the right choice as much as I possibly can, so after he lopes, I immediately withdraw and invite him in again by dropping the whip, telling him to relax, and backing up. He turns to me, stops, and faces me while licking, so I step to his side.

Again I use my hand and draw it back toward me with the invisible rope. He turns and faces me. Cool! That was the other side of his head I was looking to connect with.

His attitude is completely changed now. I walk away and he follows at my shoulder (**Figure 16**), stopping when I stop. I am now more his buddy than the other horse. That is cool because without getting him all sweaty and worn out and by doing the roundpen training correctly, we're going to have this bond forever. As long as the owner or I continue to do the right thing—always being a fair and polite leader—he will trust us.

SUMMARY

The horse learned what "Relax" means in this lesson, and he now understands that he won't be forced to submit; he will simply be invited to submit. He was able to relax because he turned leadership over to me and could trust me not to hurt him. He is beginning to understand that a kissing sound means I want him to do something and to pay attention to me. Now when I kiss and step away, the horse follows me with his head lowered and quiet.

The owner returns and strokes her horse's head. I kiss to the horse and draw him in to me. That's cool because without using any rope, I am able to draw him away from the owner he's seen every day of his

> **The horse is beginning to understand that a kissing sound means I want him to do something and to pay attention to me.**

life. Remember, I met him only fifteen minutes ago.

I go through my tactics with the owner. First, every time you approach your horse, show him the halter, leadrope, and your hand, as opposed to just walking up and putting the halter on. It's like announcing to him, "Here I am!" Also, stroke his head. If he turns away, don't reach out for his head to pull it back to you. I wouldn't like that and neither do horses. It's better to kiss, use the invisible rope, and draw him in. It's better for him to move instead of you.

The owner introduces herself correctly to the horse. When he turns his head away, she follows my advice to step back and kiss to him with her hand raised. He turns and follows her hand easily, stepping toward her. I coach her to step back deliberately and mean it—not slowly with hesitation, but quickly. Then she steps around to his other side and touches his cheek with her hand, draws it back with the invisible rope, and the horse turns immediately to follow her from that side also. Cool! There is no miscommunication here.

I coach her not to go after the horse if he starts to walk off but to draw him back with the invisible rope instead. When the owner walks on, the horse follows her with his head

When the owner walks on, the horse follows her with his head at her shoulder. She can practice this more at home later, and her bond with him will get even stronger.

at her shoulder. She can practice this more at home later, and her bond with him will get even stronger.

Our first instinct is to go after the horse if the horse moves away, but it's important to resist this urge and give the horse a chance to come to you on its own. Stroke your horse on the side of its cheek and then withdraw your hand to "pull" it in with the invisible rope. If the horse won't respond, then shoo it away to let it know it made the wrong choice, but realize it may take several times of stroking the horse and drawing it toward you to get the right response. Only then should you halter your horse.

Be consistent. If you draw your horse in a few times and all of a sudden it turns away and you chase after it, it will become a game to the horse, which isn't fun for anyone except the horse.

Here's a tip: If your horse gets ahead of you or is not leading correctly when you try to lead it with a rope, step back and away from it, raise your hand and direct it at the horse's hip (**Figure 17**), kiss to the horse to move its

hip away, draw its head gently to you with the leadrope, then reset yourself and walk off.

This lesson demonstrates why it's so important to start in the roundpen. Without the communication and understanding established there, it's so easy to give up when a horse is behaving badly and say, "This horse isn't doing what I want. I'm going to get a bigger bit," or "I just need to run this horse more." If we establish the ground rules in the roundpen where the horse learns to acknowledge us as the leader, then the horse can relax because it will understand what we want and follow our directions when we are riding.

Horses are gregarious herd animals and always searching for leaders. If we can't convey that we are the leader in a fair and easy manner, they will take control of the issue and drag us all over the place.

Total time for this session: 15 minutes

Dennis directed these exercises:

1. Trotted the horse 22 laps.
2. Loped the horse 8 laps.
3. Corrected the horse from consistently turning the wrong way and turned the horse the right way 9 times.
4. Led the horse from each side of his head without a leadrope.
5. Moved the horse's concentration from another horse to the trainer.
6. Gave the owner a lesson on how to be a leader to her horse by controlling the horse's movement with hand motions.

DRILL TEAM competitions require intense bonding, practice, and teamwork between horses and member riders. This award-winning team is unique in that it's composed all of Appalachian Purebred Gaited horses, an endangered breed. *Victoria Tollman, Equus Survival Trust, Robin Little, Rockin R Rhythm Riders*

STEP-BY-STEP

Figure 1. The horse kicks out at me with both hind legs.

Figure 2. I have to spank the horse a little with the whip.

Figure 3. The horse faces the fence, bunches his rear, and gives a little kick at me with a bounce.

THE HORSE THAT IS AGGRESSIVE

A horse with aggression issues will usually show it if given an opportunity, such as aiming a kick at you without hesitation. Some owners allow this behavior, thinking it will make them the horse's buddy, but it won't. It just tells the horse that kicking is okay and rewards the horse for doing it. Then the kicking will only get worse.

Kicking is a major safety issue. Don't settle for bad behavior in your horse just because you know how to stay out of its kicking range. Think about others. I'm a grandparent and very conscientious of children's safety around horses. Someone like my young grandson could be behind a horse like this and get seriously injured. We need to work on control issues with aggressive horses as often as it takes to get control of them. If no leadership is established and no means of communication except pain or intimidation is used, it's going to take a long time to get anywhere and someone is going to end up getting hurt. I don't take guff from horses and I don't allow aggressiveness. Instead, I establish leadership through gentle, consistent leadership in the roundpen, which fixes the problem.

The best roundpen training takes as little work as possible and it's all accomplished with proper cues, both from the horse and the trainer, and proper timing, which the trainer needs when giving his cues. The cues I'm looking for from the horse are licking, chewing, an ear on me, and hopefully the head angled away from the fence and paying attention to me. He may also show submissiveness by putting his nose down and sniffing.

THE LESSON

A gelding with aggression issues is brought into the roundpen by his owner. I ask him to lead the horse around and face me. Everything is removed from the horse's head and immediately he moves away, which indicates he wants nothing to do with us.

To control the gelding's movement, I first need some movement, so I step toward him, kiss, and raise the whip to start him trotting. After one lap, the horse stops and faces the fence, then kicks out with both hind legs (**Figure 1**) in my direction four times in rapid succession. Right away I'm seeing his aggression.

I spank the whip at his heels as he kicks out. I'm not hurting or being cruel to him, just telling him that his behavior is not okay. Every time he turns away and kicks toward me like this, he will get spanked on his heels.

I get the horse into a lope and immediately say, "Relax," while extending my free hand toward him to start turning him around. He stops and turns his rear at me. When he kicks out at me again, I spank him on his heels (**Figure 2**). He lopes again around the pen, then stops in the same spot with his rear facing me.

This horse's behavior is going to get a lot worse if I don't correct it now. I have seen this problem enough to know I can't let him throw a fit like this. He needs to know that his threats have to stop.

I make a clicking sound to the horse and move him at the trot in the other direction. He stops again by facing the fence, and when I click again to him and raise the whip, he trots off, but this time he does it without kicking out at me. He puts his head down close to the ground, which normally is a sign of submission, but it doesn't mean much when he has been kicking like that.

To control his movement, I need to move him into a lope, so I kiss to him while swinging the whip. He turns to face the fence again with his rear pointed at me and kicks toward me one time. I spank him with the whip and he lopes off. Immediately I tell him to relax because he did go into a lope.

Even though the horse is making bad decisions (kicking), it's important to respond positively to his good decisions (loping when I ask for it). I lower my hands at my side and back up toward the fence while kissing to him, which is asking him to turn toward me and then turn around. He stops, then picks up the trot and tosses his head while continuing around behind me instead of turning around.

That wasn't exactly him saying, "I want to be with you." It was more like, "If you get any closer, I'm going to knock you out!" After one lap, he stops with his rear to me and makes a little bouncing kick (**Figure 3**), but when I spank him with the whip, he turns and lopes the other way. It's not great that he kicked, but the first time he kicked four times and this time it was just a little bounce. His kicks are getting milder, and that's a good thing.

The horse trots around the roundpen, then lopes around several more laps. I tell him "Relax," kiss to him, and back up. He slows and turns to face me for the first time. Great! He looks at me and then continues on in the other direction, which is good, but when he turns, he has a look that said, "Come on, Buddy! I'm going to take you on!" I just laugh at this. Remember to have a sense of humor when working with your horse—and remember that this horse isn't mean-spirited. He just hasn't had the proper training.

I want to see a more gentle-looking horse and to see the attitude in his eye change radi-cally. Instead of saying, "Come and get me," he should be saying, "I think I like this." The horse trots a few laps before I tell him to relax. He stops this time without his rear on me (good!), but I want him to keep moving and complete the turn, not just stop, so I flick the whip and he trots off with a better attitude, although it's in the same direction. I say, "Relax," and back away closer to the fence, still trying to turn him around. My hands are down, inviting him to face me. He stops and puts his head over the fence (**Figure 4**). When I move him on, he trots with his head turned toward the fence, which is saying, "I don't want to have anything to do with you!"

Since he's not responding to my invitation to turn around by facing me, I swing my whip at him to get the lope, and when he comes to the same spot where he stopped before to kick at me, it's like a testing area to see if he will do it again. He thinks about it, but reconsiders and continues on without kicking. He lowers his head a bit, which is a good sign.

I try it once more: I kiss to get his attention and try to draw him in when he is in the same spot where he kicked out, and he stops and hesitates. He's thinking about it and though he doesn't turn around, he has given up turning his rear at me and keeps running. Cool! If we know his problem and the area where he is most likely to kick, we can fix it.

Nonetheless, I want him to turn around. As I've said before, making a horse speed up, slow down, and turn around when I ask makes him more submissive. I click to him and step back. He stops and does a little bit of chewing. He almost turns in toward me. He is saying, "I'm reluctant to give in to you just yet." As he trots on, he dips his head down toward the ground. That's submission and he does it in

Figure 4. I am close to the fence and try to turn the horse. My hands are down, inviting him to face me. This time he doesn't point his rear at me.

Figure 5. When I try to turn him, I have to use the whip to keep him going or he stops.

Figure 6. The gelding looks at me as he turns, finally doing it right.

Figure 7. The horse stops and is chewing. I extend my hand and can walk half way up to him.

Figure 8. I am able to approach and introduce myself with an extended hand.

Figure 9. After I back away, the horse turns his head toward the fence.

It never makes sense to try to control a horse with brute strength. The owner and I could try to control the horse physically with our combined strength, but even with ten more people, if the horse really didn't want to do something, it wouldn't, unless it was forced.

the same area two times. I kiss and back up, still inviting him to face me and turn around. He stops again, but I need him to keep moving to complete the turn.

We go back and forth with this: I kiss; he stops. I kiss and swing the whip (**Figure 5**) to keep him going. He goes forward, then he stops again when I kiss.

I want him to turn to face me first, not the fence, and to trot on in the other direction. I ask this of every horse I work with in the roundpen, and an aggressive horse that kicks is no exception. When I kiss to him and step back, it should cause him to turn in and face me, not the fence, then run in the other direction.

Instead, he's stopping and then running in the same direction. I am getting a little bit of his eye on me, which means he's paying attention at least. He goes into a lope and I kiss, back off, tell him to relax, and lower the whip. He slows and looks at me.

I have his eye on me and it isn't exactly the evil eye this time, which is better. I kiss again, step back, and raise a hand to invite him in. He finally turns around by facing me first (**Figure 6**)! Whew! I just laugh.

After he lopes off in the other direction, I tell him to relax and I back away again. I want to make the right thing easy for him to do and the wrong thing difficult. As he slows, he lowers his head—a sign of submission.

He has the whole roundpen to decide if he will turn or not. I kiss, raise a hand, and back away closer to the fence. He does a nice, correct turn, then hesitates, so I encourage him on and he lopes away.

His eye isn't saying now, "I'm going to clock you," as much as it is saying, "I'm almost there, Dude." Again, I kiss and back away, and again, he almost turns in toward me but stops instead of completing the turn.

I'm trying not to run him a lot and am using only light pressure so he will make the turn. I kiss and swing the whip to keep him going. He turns by facing outward at the fence. I stop him and he turns again, facing outward at the same spot where he was aggressive before. I was testing to see if he would lash out again, and he passed the test but still turned around the wrong way. I bring my arms down, kiss to him, tell him to relax, and he stops and hesitates.

He's almost there. He starts running again, but I kiss and walk toward him and he turns correctly. Cool! It's all about timing and responding immediately to his submissive body language with my own cues.

I think he just might be ready to submit, so I drop the whip, raise a hand toward him, and invite him in. He runs half a lap and stops with an ear on me (**Figure 7**). I tell him to relax and am able to walk up to him, but

Figure 10. The horse lowers his head to the ground to ignore me.

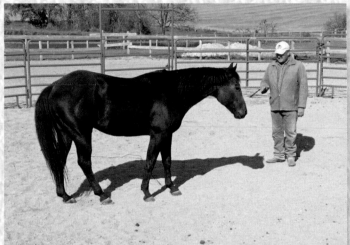

Figure 11. I motion with my hand to get the horse's head down.

Figure 12. I touch the horse's jaw lightly with my fingertips and draw my hand away. This is the procedure for using the "invisible rope."

stop halfway there. Although he is chewing, he is angled away from me, so he's not quite ready yet.

I go back for the whip and swing it. He only runs a few steps before I drop it because his eye is on me now, and it's not an aggressive eye. I raise my hand and take a step toward him while kissing, asking if it's okay to approach, but he doesn't show any reaction. I step back with my hand stretched out, and he slowly turns his head toward me.

That was an invitation from him, so I'm going to take advantage of it and step up to the horse with my hand extended toward his nose, introducing myself (**Figure 8**). He sniffs my hand gingerly and I back away to draw him toward me. That was great, but he turns his head back toward the fence instead of me (**Figure 9**); I want his attention totally on me.

I give him another chance. I approach again with my hand raised and kiss to him, but he puts his head down, ignoring me. I raise both hands in a shooing motion to get his head up, then extend a hand to him again. He turns his head toward me, inviting me in, and relaxes while I stroke his head. His eyes have changed and his poll is now even with his withers, indicating he is calm and relaxed. But then he moves forward and lowers his head to the ground (**Figure 10**), which is ignoring me. When he was running before and lowered his head, it meant he was submitting; however, lowering his head now just to "investigate" the ground and ignore me when we are standing together is rude. I won't let him do that and I make a shooing motion with my hands to bring his head up. Then I move to the other side of his head and extend my hand, inviting him in to me, but he is undecided. I extend a hand and stroke his cheek, then draw it away while stepping back. He turns and faces me, taking a step toward me while licking. Thank you! That's his first step with the "invisible rope." (See Overview, page 16.)

People assume that if they use more force with the leadrope, it will change their horse's attitude and they will get control, which I don't understand. This is control—when I can move a horse with the invisible rope. I stroke the horse's forehead and tell him to relax, then move around toward his other shoulder, again taking my time.

I don't want to turn this into a predator-versus-prey issue. The reason why horses bite, kick, or strike out is for their own survival. I want to have a positive relationship with the horse and be a trustworthy leader, not go after him, as a predator would, if he moves away.

I stroke him on the other jowl and withdraw my hand with the invisible rope, kissing to him. He puts his head down to the ground, which is the wrong reaction. I raise a hand quickly to get his head up and stroke him again on the cheek, then draw him back toward me with the invisible rope. He turns and steps toward me. Cool! I can lead him from both sides now.

He's still a little reluctant, so I motion with my hand to bring his head lower to relax him. He follows my hand down (**Figure 11**), and I stroke his forehead. Great!

SUMMARY

With some aggressive horses, it might take a little longer to establish leadership, but it's never my goal to run a horse in the roundpen until it is sweaty and breathing heavily. Anybody can work a horse until it is exhausted; eventually any horse will submit and turn toward the person if it gets tired enough.

Then it's easy for someone to think, "I won," but that's not true. The horse won't want to go in the roundpen again and won't want to be with that person if it is going to run a lot and get exhausted. That's not gaining leadership. Until the aggressive horse changes and has a nice eye, is licking, chewing, and is calm, it's too dangerous to do other training with it.

The owner comes in the pen now, and I impress on him that I don't want to proceed any further until the horse acknowledges him as the leader. I have him introduce himself to the horse by extending his hand toward the horse's nose. The horse is licking and chewing so the owner strokes his jaw and backs away with his hand extended, but the horse doesn't move. I coach the owner to mean it when he draws back with the invisible rope. I touch the horse on the jaw (**Figure 12**) and draw my hand back to show what I mean, and the horse turns toward me right away.

The owner tries again by raising his hand, lets the horse smell it, and is able to draw him away with the invisible rope. Cool! He puts the halter and leadrope on and the horse calmly accepts it. The difference in the horse's attitude is quite apparent.

I want the owner to practice leading the horse on a slack leadrope and later we can move on to other control issues. I explain about not holding the leadrope close to the horse's head. We can telegraph emotions, like fear, to the horse right through the leadrope. By holding the leadrope close to the horse's head, it means, "I don't trust you. You're going to walk off."

It never makes sense to try to control a horse with brute strength. The owner and I could try to control the horse physically with our combined strength, but even with ten more people, if the horse really didn't want to do something, it wouldn't, unless it was forced. Using only light pressure to lead a horse works the best. To do that, the rope should have lots of slack in it. Hold the rope halfway down its length with enough slack that it has a dip in it. The horse still follows easily because we have leadership now. If your horse starts walking off, use your free hand to draw the horse back with the invisible rope instead of pulling on the leadrope.

This is so important, especially with people who don't have a lot of physical strength. I encourage the owner to teach his kids and wife how to lead the horse properly. Kids particularly like to hold onto the rope close to the horse's head and yank on it. Let everyone else yank on their horses, but the owner and his family can show how to lead a horse with a slack rope.

The horse follows easily and I tell him to relax. I turn the rope over to the owner. We made some excellent progress today, but this horse is still going to need some consistent, gentle work in the roundpen to reinforce leadership before each ride so he will be safe around children.

> **Total time for this session: 13 minutes**

Dennis directed these exercises:
1. Ran the horse 24 laps, both trotting and loping both ways in the ring.
2. Corrected the horse from kicking at him.
3. Turned the horse correctly 4 times.
4. Drew the horse's head in toward him from both sides of its head with the invisible rope.
5. Lead the horse without a leadrope.
6. Coached the owner on how to lead his horse with a slack leadrope.

ON A HOT summer day, a swim on horseback, especially with friends, can be a lot of fun. This Colonial Spanish grulla gelding seems to be enjoying the pond as much as his youthful rider. *Karen Crumbie*

CHAPTER 2

CONTROL TECHNIQUES

This chapter looks at a variety of horses and control techniques. The common control issues are bringing the horse's head down for bridling, backing, giving to the bit, and stopping.

One of the tools I use is a nylon string, and I'll explain exactly how to use it to apply and release pressure. It's an easy, light method that is noninvasive, and there is no pain or intimidation associated with it. I also use verbal cues such as what I used in the roundpen to tell a horse to relax.

Last but not least, I use the reward system. When he gives the right response, I reward the horse with small amounts of grain, the same thing horses eat everyday. I call it a treat, but it's not the sweet kind of treat. With it, you're not only getting the desired response from your horse in a quick and efficient manner, but you're building a bond that is unbelievable.

I like to give grain instead of another treat because it's easier to feed a small amount at a time. I only use about a tablespoonful each time to reward the horse until it has learned the exercise, then I gradually phase it out.

Rewarding with treats goes against the warnings many of us heard because we were told it would cause a horse to bite. The method I use prevents a horse from biting, and instead it causes a positive and quick response, usually every time. I don't use it as bribery; it's a reward. I'm actually paying the horse for what it is doing, which is positive reinforcement. I train by using light pressure and only give a small handful of grain right away if the horse does what I ask. Horses understand rewards, and this method allows them to get into a relaxed mode that supersedes any other negative attitude they may have built up against training. They learn quicker, retain the lesson longer, and look forward to training. With my method, they don't get pushy or bite. They ask for a reward and take it politely.

The most important thing is for you to remain consistent with the system. The horse

It's so important to learn the fundamental control techniques before moving on to more advanced lessons with your horse. *Nick Vedros, MindFire Communications*

57

IT'S EXHILARATING to ride horses at a full gallop along a sandbar, splashing through the water. These Marsh Tacky horses trust their riders and are easy to handle in any terrain, but especially in water. *Dwain Snyder*

will understand better if you approach it every time in a calm, kind manner. Anger has no role in training. Reward your horse for a proper response and always quit on a good note. When you do this appropriately, you'll have a lifelong bond with your horse.

And remember, before you attempt any control techniques, establish leadership in the roundpen first. It's imperative that we have some line of communication and understanding before we move on. Your goal is always to have a positive, healthy, and trustworthy relationship with your horse, which is the end result of good training.

HEAD DOWN

Some horses raise their heads every time a halter or bridle is put on them. If the owner tries to reach up over the horse's head to get its head down, the horse resists and throws its head up. It will really help if the horse learns to put its head down on command. When I am on a trail ride, it's great when I can just tell my horse to bring his head down and he does it.

Today I'm working with a gelding that has this problem of raising his head. We established leadership with the horse previously in the roundpen, and we quit on a positive note, which was allowing him to stand when he responded correctly. So I'm ready to fix the problem of him raising his head.

THE LESSON

I stand beside the gelding holding his halter and leadrope with a bucket of grain beside me. When teaching this lesson to your horse, stand at his side for safety because standing in front of him is dangerous. This is also where we stand when we put the halter or bridle on, so he should be used to it.

Only use about a tablespoon of grain each time you reward your horse.

I offer him a handful of grain with a closed hand, and he tugs on the leadrope while nibbling on my hand with his lips, which is rude. So I'm going to "pop" or bump him on the mouth with the closed fist (**Figure 1**) that holds the grain (palm up) just enough to cause him to jerk his head up. I continue holding the grain in front of him, waiting for him to ask politely. He touches my closed fist lightly with his lips, which is asking nicely, so I give it to him. He understands now how to ask for it instead of grabbing.

Now that the gelding knows how to take the grain politely, I can start teaching him to bring his head down. I don't like to reach over a horse's head to bring it down for bridling or haltering. Most horses resist a hand or an arm over the top of their head by throwing it up higher or moving it out of reach (**Figure 2**).

I use the verbal command "Head down," pull lightly on the halter, and offer grain.

Figure 1. I pop the horse's mouth with my closed fist just enough to make him raise his head.

Figure 2. The horse resists when I reach to put a hand over the top of his head.

Figure 3. I motion with my hand to bring the horse's head up and lift his head with the other hand.

Figure 4. I use a hand motion to bring the horse's head down.

Figure 5. When the horse brings his head lower, he will get rewarded.

STEP-BY-STEP

Figure 6. I use only a thumb and forefinger to give a very light pull on the halter.

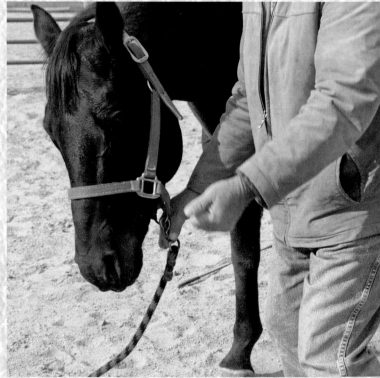

Figure 7. A very light touch and a reward are enough to bring the horse's head down.

Figure 8. I rest my arm on the top of the horse's head with my hand on the other side and offer him a closed handful of grain to bring his head down.

Figure 9. I push gently on the opposite side of the horse's poll to bring it in toward me and reward him.

Figure 10. I use an upward motion, and the horse brings his head up.

If the horse holds up his head because he is nervous, I signal to him with an extended hand, and say, "Relax." He knows that word from the roundpen work we did before.

When the horse brings his head down to get the grain, I give it to him as a reward and release the pressure.

This training method also controls movement by teaching a horse to both lift and lower its head to a comfortable level. When I tell the horse "Head up" or simply "Up" while raising my free hand in an upward gesture (**Figure 3**) and gently lifting my other hand on the halter, the horse raises his head and gets rewarded with a bit of grain.

Now that his head is up, I let him think about it a minute, then tell him "Head down," and apply a slight downward pressure on the halter together with the other hand gesturing downward. The horse brings his head down (**Figure 4**) to get some grain. When I repeat "Head down" and he brings his head down lower (**Figure 5**), he gets rewarded with grain again. If he raises it up without my asking, there is no reward.

It's a simple lesson: When I say, "Head up," and the horse raises his head, he is rewarded immediately with grain. When I say, "Head down," with a light pull on the halter, barely touching it, and the horse drops his head down, he is handed grain. It's all timing. The horse learns to raise and lower his head when he is told if he is rewarded each time until he knows the commands. This exercise is another opportunity to control movement and establish leadership. The more we do this, the better off we are, no matter how old the horse is.

Do you know what lightness means? I love my horses to be light, which means I can ride them with cues that are almost invisible. The cue to bring a horse's head down takes only a thumb and forefinger (**Figure 6**) giving a light, steady pull on the halter plus the verbal command. See how light you can get with that pressure. This will help your riding too.

If the horse holds up his head because he is nervous, I signal to him with an extended hand, and say, "Relax." He knows that word from the roundpen work we did before. Then I say, "Head down," while using light pressure on the halter, and he brings his head down (**Figure 7**). I give him a reward and a little time to let him think about it.

The cue to a horse to put its head down is just a couple of fingers on the halter. Now I want to transfer that cue to the top of his head where he is haltered and bridled. Usually a horse will throw its head up if we reach for his poll, often resisting being handled there. So I say, "Head down," while I rest my arm on the top of his head behind his ears and with my hand on the other side of his head. If he doesn't bring his head down, there is no treat. I say again, "Head down," offer him a closed handful of grain, and the horse lowers his head, asking for it. If I continue to do this, he will learn that when I put an arm over the top of his head, that is the cue to bring his head down (**Figure 8**), and he will get a treat.

Because the horse I am working with is so resistant, we don't want to push his head to the ground, so we'll go inch by inch. We'll ask him to lower his head a little bit more each time we do this and end up getting a lot more cooperation from him.

If I put my hand over the top of his head and he lowers his head all the way to the ground, I don't want that either. I say, "Head up," put my hand under the horse's jowl, and lift his head with a light touch.

I can turn his head in toward me with my hand on the opposite side of his poll by pushing it toward me with another light touch (**Figure 9**). This is where my hand will be when I bridle the horse. I say, "Head down," and he brings his head down to a relaxed level that is more ideal for haltering or bridling. When you do this, be safe and stand way to the side of the horse's head.

If he raises his head instead of lowering it, I don't give him a treat and, more important, I don't fight with him. I say, "Relax," which he knows from our roundpen work, and ask him to lower his head just a little bit. I put my arm over the top of his head again, say, "Head down," offer him grain in the other hand, and

he lowers his head to get it. When I say, "Up," and use an upward hand motion, he brings his head up (**Figure 10**).

SUMMARY

This training is something to work on and have fun with. All it takes is just a couple of minutes each day. Many owners want to be friends with their horses, so they give them a treat whenever they ask for it; that's what teaches a horse to be pushy. The goal here is to always be in control, and an integral part of that is rewarding with the right timing and only when the horse earned it.

Eventually the lightest touch on the halter or just the verbal command will bring the horse's head up or down. It sure beats jerking on the horse, it's more fun for both of you, and you build a bond that's unbelievable.

> **Total time for this session: 5 minutes**

Dennis directed these exercises:
1. Taught the horse how to take treats (grain).
2. Taught the horse to lift and lower its head on cue.
3. Taught the horse to bring its head down when Dennis put an arm over its head.
4. Taught the horse the verbal commands to bring its head up and down.

DID YOU KNOW?

COMPETITIVE DRIVING requires special teamwork and bonding between driver, navigator, and team. These Gotland Ponies—a rare breed from Sweden—and cohorts are enjoying a fast-paced cross-country marathon practice. *Victoria Tollman, Equus Survival Trust*

BACKING

It's important that horses know how to back up. When a horse throws its head, backs up crooked, or fights the bit, it's no fun for the rider. Good backing is especially important in an event like a pleasure class, where the judge looks at each horse backing up.

Before working with a horse on backing, it should have already been through round-penning several times. The issue of leadership has already been settled, partnership has begun, some form of communication is in place, and bonding has started.

The horse should also know how to take grain correctly from someone's hand when it is offered as a reward. To do that, have a bucket of grain handy and offer the horse a handful of it. If the horse is aggressive or pushy about taking the grain, pop (bump) the horse once or twice on its mouth with the hand holding the grain and wait until it asks politely. This doesn't hurt the horse, but it helps it understand how to take the grain before the backing lesson can begin.

The horse should have a bridle on for this lesson and the bit should not be harsh. Pressure is used on the bit, which is released when the horse backs up after it has been asked. Don't use a curb bit because it causes a measure of pain and intimidation. Instead use a snaffle bit, which is light, fun to work with, and causes no pain. The horse will respond well to it.

In using the snaffle, I recommend that the O ring (or D ring) fit snug enough to make two or two-and-a-half wrinkles in the corner of the horse's mouth (**Figure 1**), so it fits where the bar of the mouth is, where there are no teeth. If you have to use a curb bit (a straight bit would be best), there should be one or one-and-a-half wrinkles in the corner of the horse's mouth.

The horse we are working with today doesn't like to back up. He throws his head up, resists the bit, and backs up crookedly. We'll give his backing lesson in five parts.

LESSON, PART 1

Standing in front of the horse, hold both reins with one hand under the horse's neck and the other hand holding some grain. Show the horse the grain without giving it to the horse and then hold it low and halfway to the horse's chest. Offering the grain lower than the horse's head will bring its head down. Tell the horse "Back" while lightly pulling the reins back toward the horse's chest. When the horse takes several steps back without fighting the reins, quit pulling and give it the grain. The horse should step back easily with its head low and not tossing because it will be reaching for the grain.

The steps are listed below:

1. Show the horse the grain in a closed hand under his nose so there is no mistaking it is there.

2. Bring the grain back closer to the horse's chest area.

3. Say, "Back," to the horse while simultaneously pulling lightly back on the reins with your other hand positioned under the horse's neck.

4. Give the grain (**Figure 2**) and release the rein pressure when the horse has its head down and has backed up without fighting the bit.

Only when the horse steps back with its head lowered and isn't resisting should the horse be rewarded with grain. Give the grain to the horse immediately and release the pressure on the reins when the horse backs up correctly, even if it is only a step or two.

The grain is only a handful, and the pressure on the reins should be released right

Figure 1. There are two wrinkles in the corner of the horse's mouth using a snaffle bit.

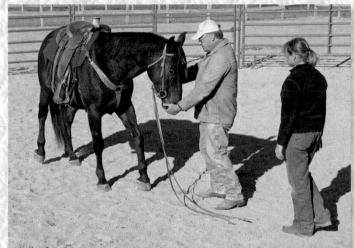

Figure 2. I give the grain after the horse has backed up. The owner stands by to help with the next part.

Figure 3. I hold the grain low so when the rider asks him to back up, he will have to reach down and back for it.

away. Timing is important. Don't give the grain too soon if the horse is fidgeting or too late after it has focused on something else.

LESSON, PART 2

Have another person help with the second part of this lesson. I love this part because it's always fun to train with partners. A second person mounts the horse while the first person gets ready to treat the horse with a hand under the horse's neck (**Figure 3**).

The rider should only ask the horse to back up when its ears are on her. That means the rider has the horse's attention. Kissing to the horse will get the horse's attention. So will playing lightly with the reins, gently bumping them, or lifting them. When the horse is ready and attentive, she lightly pulls back on the reins (**Figure 4**) and tells the horse "Back." When the horse takes a step or two back and lowers its head, the reins should go slack. That is when the rider quits pulling and the first person treats the horse.

There is no need to pull hard on the reins because the horse wants to lower its head and back up for the grain. The reins should get loose when the horse brings its head down. It's important to find that moment when this happens to reward the horse with grain and lower the reins. Timing is essential.

The rider doesn't release pressure on the reins if the horse's head is up or it will teach the horse to bring its head up for backing. Even if the horse's head is level with the withers or its mouth is open and resisting the bit, don't release pressure on the reins. When the horse takes a step back and lowers its head, then the rider immediately releases the hold on the reins. Then the horse gets a reward of grain at the same time from the other person.

> # There is no need to pull hard on the reins because the horse wants to lower its head and back up for the grain.

LESSON, PART 3

The third part of the lesson slightly modifies the method. Now the treat-giver holds the grain close to himself—instead of under the horse's neck—and waits for the rider to pull the reins and say to the horse, "Back." The pressure on the reins remains light. (I can't stress this enough!) The rider remains relaxed and projects that to the horse. When the horse brings its head down and backs up, the rider releases the reins and only then is the horse offered grain (**Figure 5**).

The rider needs to be careful not to jerk on the reins. Just lift them lightly. If the horse doesn't respond right away, the reins should be slowly pulled back and planted tight against the rider's thighs to hold them steady until the horse backs up. This keeps the reins in place, and the horse will automatically get bopped in the mouth if the horse tosses its head.

The horse's head should come very low, about halfway to its chest to get its reward. It should be given to the horse low and close to its chest so the horse will have to step back and tuck its head in to reach it. If the horse cranks its head to the side to reach the person with the grain, it can be offered by reaching around (**Figure 6**) to the other side of the horse's head.

Figure 4. The rider pulls the reins to back the horse. He lowers his head to get the grain.

Figure 5. Only after the horse backs up will I give him the grain.

Figure 6. I reach around to the other side of the horse's head to give the grain so he will tuck his head properly.

Figure 7. I wait a few steps away until the horse backs up and quits pulling to give him the grain.

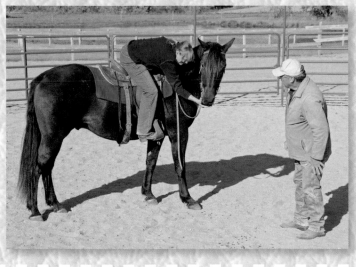

Figure 8. The rider reaches down to give the horse a treat from the saddle.

Figure 9. The proper hand position to back up the horse: Both hands are lifted a little and held wide apart. The reins are crossed in front of the saddle.

Figure 10. The rider holds the reins low and against her thighs until the horse gives to the pressure.

Figure 11. The rider wants the horse to lower his head more, so she doesn't release the pressure.

LESSON, PART 4

The fourth part of the lesson begins with the treat person taking a few steps away (**Figure 7**). The rider pulls back on the reins very lightly at first and then slowly down and back while telling the horse, "Back." The rider's hands shouldn't be too high. When the horse takes one step back with its head down, the rider releases the pressure on the reins and the treat person steps up to give the horse a grain reward.

Soon the horse will start to respond without any head lifting or shaking. Its head will come down and the horse will step back smoothly and straight because the grain is its reward. If the horse takes one step back without turning to one side or the other, that is the moment to look for. The rider should be able to feel the release in the reins when this happens because the reins will go slack when the horse puts its head down. That is when the rider quits pulling and the treat person rewards the horse. It's the same lesson as before, except now the treat person is a few steps away, so the horse learns to follow the command without the treat right under its nose.

LESSON, PART 5

The final part of the backing lesson is for the rider to reward the horse from her position on the horse's back. The rider can have the grain in a horn bag (a small saddlebag on the horn or pommel of the saddle) to give a treat while mounted. The rider tells the horse to back, backs the horse as before, and waits for the release in the reins when the horse brings its head down and steps back. Then she releases the hold on the reins and reaches down to give the horse a treat (**Figure 8**). There is no longer a need for another person to help. If the horse lifts its head, it gets no treat. If the horse steps back without fighting the bit or backing crooked, it gets a treat.

When the rider picks up the reins to back the horse, she gradually lifts them to signal to the horse that backing is requested. She can finger them or play with them if the rider doesn't have the horse's attention. Then she uses a gradual squeeze, pulling the reins back lightly at first to ask the horse to back up. It's best to use both hands simultaneously and evenly in a steady pull that starts out light and releases when the horse takes a step back.

The rider's hands should be low, close to the pommel, but wide enough apart to prevent the horse's head from going up. I like the reins to be crossed in front of the saddle and a rein held in each hand (**Figure 9**). That way they can be pulled back evenly and released quickly. They should be the same length and pulled back with the same pressure.

The reins should not be jerked. If the horse doesn't back up, the reins should be pulled back slowly and held steady against the rider's upper thighs to bring the horse's head down. If the horse tosses its head or moves around, the horse will realize it is causing pressure on its mouth by pulling against the rider's unmovable hold. The horse will learn to relieve the pressure by lowering its head and backing up, so the rider shouldn't worry about it or move her hands until the horse lowers its head.

When the horse backs and the rider releases the hold on the reins, it's good to literally throw them down on the horse's neck instead of holding them so that there is no mistake the pressure is off. If the reins are crossed, they won't slide off the horse's neck and can be picked up easily again.

We want the horse to back up with very light rein tension, so start slow and don't crank on the horse's mouth. Pull back low to get the horse's nose down and plant your hands on your thighs (**Figure 10**). The instant the horse releases pressure on the reins (when it tucks its head in and moves back), that is the signal to throw the reins down. If the horse continues to fight when asked to back up, especially after it has done it right a few times and been rewarded, don't be intimidated and let the reins go. If the horse only tucks its head in but doesn't move back, don't release the rein pressure. Hold on until the horse takes a step or two back or even just brings its head down a little more. The horse may surprise you and tuck its head and move back several steps with no tugging. Then instantly release the reins because the horse quit fighting. The bit we're using is easy on the horse's mouth and there is no pain associated with backing, so it shouldn't be resisting it.

When the horse is responding well, you can quit giving grain as a reward; dropping the reins should be reward enough.

Lift the reins again and squeeze them lightly. If the horse backs up several steps well with its head down, drop the reins on the neck right away. If you make a mistake and don't drop the reins at that moment, the horse won't forget it, but it will forgive you. The nice thing is when we make a mistake, there is no pain. We can keep working at it. It doesn't matter if the horse backs one inch to start with or three feet, but try to get a little more each time. Eventually the horse will step back right away if you consistently and immediately release the pressure at the right time.

Many owners want their horses to back up a lot the first time they teach this lesson, but that's not the way it works. Instead of trying to get numerous steps at one time, reward the horse if it takes only one or two steps back. That's enough for a reward at this early stage. Even if the horse takes a small step back or a nervous horse just tries to relax, as long as its head is down, reward it. Once the horse understands, you can work on increasing the number of steps back.

When you ask the horse to back and it does it right, but you want its head down a bit more, don't release pressure right away until the horse takes the tightness out of the reins. If it backs one step correctly, but you feel it can do two, ask for two perfect steps back with the horse's head down. If the horse tucks its head in, stays relaxed, doesn't bite or fight the bit, and steps back two steps, then drop the reins. It won't win any points in the show ring just yet, but if your horse does it, the horse is getting better.

If the horse backs up right away when you pick up the reins, but you want its head down lower, wait until the horse lowers it more (**Figure 11**) and then drop the reins as a reward. You can feel in the saddle when the horse puts its head down: Its back elevates and rolls up a little and the horse's rear end will come underneath it, which is actually more comfortable for the horse.

Ask for a little more each time the horse does it right. In the beginning, we only wanted the horse to try; now we want it to back until we are ready to quit. Try getting the horse to back up evenly more than a few steps with its head down. If the horse does it, that's great, and it's time to drop the reins and quit.

Take your time; there is no hurry. We want the horse to respond to the lightest pressure possible on its mouth.

Figure 12. The horse's head is up, fighting the reins.

Figure 13. The horse tosses his head, resisting the bit as he backs.

Figure 14. The horse is starting to flex his neck and back up correctly.

STEP-BY-STEP

Figure 15. The horse has begun tucking its head in and backing up straight, but he needs to drop his head more.

Figure 16. The horse's poll is relaxed and it is backing with just a lift of the reins. It's not perfect, but it's better.

Figure 17. The horse starts to back up with just the reins being lifted without pulling on them. The horse will get better at lowering his head with more practice.

PROBLEM SITUATIONS

A horse's educational chart normally looks like this: It starts where we don't want it to be, then it goes up (sometimes rapidly), but then it comes back down. It's going to be perfect a couple of times and then the horse will usually start fighting the bit again. It's normal for the learning curve to get better and then revert and go back down for a while, but the next time it may be the same or get better if we're consistent. So keep trying.

It also takes the right timing to succeed. If the horse is allowed to move its head around or mouth the bit and the rider quits the lesson there, that's the kind of movement it will do when asked to back in the future. If the horse backs but is still resistant—lifting its head, fighting the bit, or backing crooked—and the rider gives in, that's rewarding the horse and it will keep doing it, so it's best not to quit too soon. Horses back up to release the pressure, and the only time to release it is when the horse is doing it right.

If the horse is not being attentive, you can tap your fingers on the reins to get its attention or kiss to the horse, but wait until it is paying attention to repeat the lesson. This is important because the horse can only concentrate when it is ready. If you start asking the horse before it has its ears on you, you've got a fight coming. If the horse is mouthing the bit or looking somewhere else, it is not ready, so don't even try. When its head is quiet, that's when the horse is ready.

In the case of a horse that is stiff necked and won't bring its head down (**Figure 12**), be consistent and keep asking the horse to back up until its head goes down. The moment the horse backs correctly with its head down and giving to the bit, it can be felt in the reins and the rider should respond

quickly and drop the reins as a reward. Even if the horse's head goes down an inch, the reins can be released because it tells the horse that's what you want.

Don't ask the horse to back when it's throwing its head up. Use your pinky finger to squeeze the reins for light pressure to bring its head down. If the horse lifts its head a little but then brings it down and steps back once, drop the reins. It doesn't look pretty, but accept it for now. It's so important to quit there to reward the horse when the pressure is released on the reins. It's up to you to reward at the right time.

If you are doing everything correctly and the horse tugs a little on the reins when you ask it to back, it means the horse is still fighting it. Don't release the reins until the horse does it right. It's the same if the horse braces its head against the reins or tosses its head (**Figure 13**) as it steps back. Don't release the pressure until the horse steps back again and brings its head down. That's the moment to look for and release the reins.

You also might have a situation where you've been getting a couple of good backups and then you get one where the horse fights because your timing is off for dropping the reins. Otherwise the horse might just be trying to get away with it, or it could be a result of the learning pattern I talked about before: The horse does great, then reverts back to being bad or indifferent before it gets better again. You might have to retrain the horse for ten minutes to get the same result, but don't give up. Be patient and try again.

If the horse lowers its head and starts doing one or two decent steps back again, drop the reins on its neck. If the horse fidgets or reverts to bopping its head or you only get one good step back from it, accept what the horse gives as long as it backs up correctly, but keep asking for a little more each time.

If the horse's head goes up, say, "Head down," and wait until the horse takes several steps back with its head down before releasing the reins. (The horse will know what this means if it had the lesson for bringing its head down from pages 59–65.) Or you can tell the horse to relax, which he should have learned in the roundpen. Don't fight with him.

If your hands are planted on the pommel and the horse tosses its head, let it bop against the reins so the horse will quit. The horse will learn you're not doing it; the horse is. The horse may take several steps back and toss its head, then take more steps back and toss its head again before it tucks its head and does it right. When the horse backs correctly (**Figure 14**), quit and drop the reins.

Timing is important. The moment to release the pressure on the reins has to be felt by the rider. That moment is when the horse has tucked its head and taken a step back without its head coming up. Now that we're at the end of the lesson, if the horse backs up with only a slight tug, but it wasn't until the third step that it tucked its head, wait until then to release the reins.

SUMMARY

It sure isn't pretty when we start working on backing and the horse fights against it, but understand the concept: Start with the lightest pressure possible on the reins, and if the horse responds correctly, release it. Learning how to ride with light hands is imperative. Too harsh a bit or heavy hands can cause head tossing or fidgeting, and 90 percent of the time are the problems with backing up.

There are several steps to backing up correctly. In the final stage, the rider gets the horse's attention and lifts the reins. If the horse backs up with relaxed flexion in its neck, that is what we want. Don't release the reins if the horse's head is higher than its withers. Bumping the reins a bit with your fingers will help bring the horse's head down. Also, don't drop the reins if you can feel the horse cranking on the bit or fighting it, even if its head is down. (**Figure 15**). You want the horse to tuck its head in. The second the horse does that, drop the reins on its neck.

After the horse can do a step or two back with its head tucked in and no fidgeting, ask for more steps until the horse is backing up adequately with a light touch on the reins. Work at it until you can ask for more controlled movement, not only with backing, but also with getting that nose in and the poll relaxed (**Figure 16**) and then put it all together.

The more you practice this, the more confident and better your horse will get. (**Figure 17**). When the horse starts responding as you pick up the reins, that is when you will know the message is getting across. Eventually you can just lift the reins and the horse will back up nicely with its head down. That's the time to quit because you don't want the horse to get bored. Quitting when the horse is doing it right and still interested is so important. Always quit on a good note.

Total time for this session: 34 minutes

These are the exercises Dennis performed:

1. The horse was corrected from lifting, tossing, and cranking his head, mouthing the bit, backing crookedly, and bracing against the bit when asked to back.
2. The horse learned to back up when the reins were lifted a little.

A WELL-MANNERED pleasure horse should be rode with a loose rein, have a bend at the poll, and carry its head vertically. Here an owner enjoys her beautiful Arabian—a Canadian National Top Ten Western Pleasure Jr. Horse.
Bridget Lockridge, Equine Photo

GIVING TO THE BIT

Each time I touch the reins to turn a horse, it should respond right away by turning without any objections. This is called "giving to the bit." The softer the horse's response and the more it gives to pressure, the more control I have. Riding a horse that isn't soft is like driving a semi with no brakes and no steering. I don't want to go anywhere if I have a horse that isn't soft and that fights the bit.

Some horses fight the bit so badly that they are out of control. Bracing against the bit, stretching out or tossing their heads to pull the reins out from their riders' hands or being inflexible for turning are all little things that add up over time and make riding miserable. I want the horse to give to the bit with just a little squeeze on one rein. That squeeze should be all it takes for the horse to turn.

I've heard people comment that they have to be careful with their horse because it is soft-mouthed, which means that it has a sensitive mouth—or maybe it's already learned to give to the bit and the owner didn't recognize it. Other horses are classified as hard-mouthed, which means they press against the bit and don't give to it. Horses like this can still be trained to be soft.

I don't like seesawing on the reins (bringing the horse's head way to one side and then the other), and I don't like cranking a horse's head around. I make a horse soft by pulling on one rein and taking my time. I don't give treats for this training. The horse's reward is getting away from the pressure of one rein pulled back.

The horse needs to be relaxed to get soft, which means the horse doesn't feel any pain. So it's best to use a snaffle bit instead of a curb bit, which can be fairly harsh. Snaffle bits apply and release pressure without pain.

THE LESSON

We start by asking the horse for a small turn by holding one rein solidly in one spot. When the horse turns, the rein tension is released. This simple procedure produces softness, which is what we want. Eventually we can just squeeze one rein and the horse will turn right away (**Figure 1**). This is called giving to the bit or getting the horse "soft"; both mean the same thing.

To teach this to a horse effectively, we first need to control its movement, which is establishing leadership. So start by making the horse walk while you are mounted in a quiet area like the roundpen. Ask the horse for a small turn as it walks by squeezing one rein lightly to the side. The horse must be moving when you squeeze the rein (**Figure 2**). Then hold the rein tight against your thigh or saddle and the horse will incline its head to that side. When the horse does this, it is releasing the pressure, not you. Reward the turn by dropping the reins on the horse's neck. If it turns well without fighting the bit, increase it an inch more the next time.

You can either turn the horse in a small circle or just bend its head a little to the side. It doesn't matter where the horse is going; it's immaterial. We are not guiding the horse; we are just practicing turning. The important thing is to get forward movement like we did in the roundpen lesson.

When you squeeze one rein to the side, you should get some response. If all you get is the horse turning its head only a little bit, even if it's just an inch, drop the reins on its neck to release the pressure. That little bit is good enough to begin with.

If the horse bumps its mouth by tugging on the bit or bracing against the rein with its nose up, the horse will find that it

Figure 1. The rider only needs to lightly squeeze one rein and the horse responds by turning. Notice the slack in the rein.

Figure 2. The horse must be walking while the rider squeezes one rein to the side.

Figure 3. Keep the horse walking while holding one rein tightly against the pommel until it turns.

Figure 4. I direct the rider to throw the reins down so there is no mistake that the pressure is off. The reins should be crossed on the horse's neck so they can be easily picked up again.

Figure 5. The reins are held lightly with just a couple of fingers.

Figure 6. The rider is able to turn the horse on the other side with just a light touch, which is the horse's soft response.

doesn't like that. If you maintain your steady hold on the rein, the horse will eventually bring its head around. It's the horse's fault if it keeps bumping or tugging against the bit. Hold steady, and make it the horse's idea to bring its head around. You can actually feel it when the horse bumps its head on the bit and also when it releases tension on the rein as it turns. You don't want to miss the opportunity, so concentrate on what the horse is doing. When the horse turns its head, throw down the reins on its neck to be sure the pressure is released right away. The horse will understand this as its reward if the timing is right.

Then pick up the reins again, move the horse forward at a walk, lift the same rein (lightly at first), then pull it back just a little; if the horse doesn't turn right away, plant that hand either on your upper thigh or the saddle pommel. If the horse turns its head to that side to release the pressure, it is giving to the bit correctly, and you should drop both reins on the horse's neck as a reward.

If the horse is turning well without fighting the bit, ask for a little more each time. Walk the horse again and lift the same rein, squeeze it back, and plant your hand on the pommel with the rein an inch shorter than before, so the horse will turn its head while walking. Ask for an inch more each time and release the pressure when the horse responds correctly by turning. When you can turn the horse with one rein, you have established control.

Do the exercise repeatedly until the horse turns well: Pull the same rein back to the side of the saddle, hold it steadily against the pommel (**Figure 3**) or your thigh, and keep holding it there until it turns. This works better than pulling the horse's head tightly around, but keep the horse moving forward

by clucking to it and give enough slack in the other rein so it can turn.

If the horse responds by bumping repeatedly against the bit when you turn it, the horse will be pressuring itself if your hand is planted firmly and braced against it. You are not doing that to the horse's mouth. It is as if the horse is hitting its head against a wall. Pretty soon the horse will think, "I don't want to do that anymore!" and quit.

If the horse does some initial good turns in response to a light touch on one rein, but reverts back and turns its head only a little, then drive the horse forward with your hand still holding the rein in place until the horse bends its head well again. Then drop the reins immediately on the horse's neck (**Figure 4**), which takes good timing. Dropping the reins must be done right away, the second the horse gives to the bit.

Once you get a good response a few times when the horse brings its head around, you can go further and pull the rein back more. The horse may bump against it a time or two to test it, but it will eventually lower its head and turn. That's when you release the pressure.

After you have the horse bending its neck and turning on one side, walk the horse again and bend its neck on the other side the same way. Lift that rein, and hold it gingerly—lightly at first—with only two fingers (**Figure 5**).

If the horse doesn't turn immediately, then plant your hand, bracing it on the saddle pommel. If the horse is not giving to the bit just yet, wait until it turns its head and then release the rein. If the horse raises his head, keep holding the rein there and wait until the moment when it flexes its neck and turns its head, then instantly drop the reins.

Figure 7. The rider turns the horse's head using only a thumb and one or two other fingers, then drops the reins on the horse's neck.

Figure 8. Get the horse turning its head enough to almost touch your leg or foot. Be sure to leave enough rein on the off side to allow the horse to turn.

Figure 9. When the rider turns the horse on the other side, the horse bends enough to almost touch the rider's foot.

You can shorten the rein length inch by inch each time the horse does it right until it is almost reaching your stirrup. Eventually you can just lift a rein and get softness (**Figure 6**).

This is training a horse for a positive response, which means the horse will want to get away from that pressure without overreacting or fighting it. Never jerk on the reins. Just calmly lift one rein to the side and when the horse turns its head, drop both reins. It's a control technique, and control is what we want.

The other technique out there, which I abhor, is when the poor horse is put out in the roundpen or arena with its head tied around by one rein to the cinch ring on the side of the saddle and just left there to turn in circles for hours. It's the old way of breaking a horse to turn its head, but it's not training. It's pure laziness. And I think it's animal abuse. I really do!

My method is no abuse, when we can lift one rein gently with two fingers (**Figure 7**), letting the horse thump or tug on it if it wants to until the horse bends its head. There's no pain or intimidation.

When your horse is turning well on the other side, go back to the first side. The horse may test this side again like it did with the other one by bumping a time or two, but you're not doing it to the horse or jerking it either. Eventually when the horse bumps enough, it will turn its head.

One side is going to come quicker than the other. No matter if it's horse or human, there is a tendency to use one side more than the other, so you're going to get one side of the horse remarkably soft. When that happens, work on the other side and make that one even better. It's a goal for you and fun to do. If the horse cranks its head up or tries to stretch

the reins out, but then lowers its head and bends it around, drop the reins to let the horse know it made the right choice.

When you get that side soft, go back again to the first side until the horse bends easily on both sides. Eventually, the horse will get so soft on the bit it will turn both ways with no bumping. The horse will feel the slightest touch on the rein and you can lift either one to turn it. Be sure to drop the rein right away when the horse does this and drive it forward again. It takes timing; everything is immediate.

As you practice, try to get that extra inch back whenever you lift a rein to turn your horse. If you do it right and hold back a little more rein each time, the horse will eventually turn its head almost enough to touch your leg (**Figure 8**). It may even do it on its own when you just touch the reins. Then you know the horse is getting the message.

Don't be afraid to make a mistake. Take your time and don't get discouraged. Both you and the horse should be relaxed. We want a light, quick response where the horse turns its head a lot, and we will get it if we keep working on it. Pretty soon the horse will turn its head easily when you just pick up a rein; as soon as you touch it, the horse is going to bend, knowing that will relieve any pressure. You may be surprised at how much control this means.

Let your horse turn in a small circle instead of a gradual turn to the side, and if it completes the circle with just a light touch on one rein, that is getting fairly decent. Do it at a walk continuously until your horse gets really soft, when you can lift a rein and it turns its head instantly. Get it to that point before you go to the trot or the lope.

As the horse gives quicker to pressure, it will start to relax and bend at the poll (**Figure 9**), which is more comfortable for it. This gives you total control, which is great!

SUMMARY

The more you practice this lesson with your horse, the more fun you're going to have later when you're out on a ride because of the immediate response you'll get from your horse. It's cool when you can lightly lift one rein and your horse easily flexes and turns. You can't practice it too much. It's not ridiculous to say you can practice it hundreds of times.

It's fun for me when I'm riding out in the mountains and I play around with using one rein to turn my horse. You can't have too much control or a horse too soft.

While you're trail riding or warming up in the arena for a show, or whatever you're doing with your horse, just work on it.

Stay with it and have fun.

> **Total time for this session: 9 minutes**

These are the exercises that Dennis performed:

1. When the horse was reined to make a turn, it was corrected from bracing against the bit, stretching its head out, opening its mouth, tugging or bumping on the reins, and otherwise resisting turning its head.

2. The horse was taught to turn easily with just a lift of one rein. The horse turned with a soft, quick response and turned its head almost enough to touch the rider's foot.

COLONIAL SPANISH Horses can be great at speed events. During the Back-Firing the Prairie Class at an Indian Horse show, a gelding flies out of the arena after crossing a brush pile while his rider dispatched a simulated torch. *Karen Crumbie*

Figure 1. The cue to stop is to lean back in the saddle with feet pushed forward.

Figure 2. The reins should be kept loose when giving the cues to stop.

Figure 3. The rider's hands flex down quickly at the wrists to "pop" the horse as a correction.

STOPPING

It's important to have complete control while in the saddle, whether it means the horse neck reins, gives to the bit, backs up when you ask it to, or anything else needed for riding. Having brakes on a horse is one of those vital elements of control, so we'll work on stopping next.

For this lesson, it's best to practice first in a quiet area like a roundpen, without a lot of distractions going on. The wrong time to train a horse to stop is when other horses are coming and going and your horse wants to be with them. The rider will be more tempted to pull hard on the reins to get control, and we don't want that. Later, when the horse has had some good responses with stopping, it will be okay to practice with other horses around.

We don't want to use pain to stop a horse, so it's best to use a snaffle bit. A curb bit is too harsh. I also have the rider carry grain in a horn bag on the saddle to reward the horse.

THE LESSON

I teach this lesson in baby steps, which means we start training the horse to stop from a walk before we even consider progressing to a trot or a lope. The horse must understand and respond well at the walk before it will ever be able to do it at a faster gait.

To begin the lesson, ride the horse forward at a normal walk. When you want the horse to stop, lean back in the saddle and push your legs and feet forward in the stirrups and up and out to the sides. Do it radically at first so the horse understands the cue (**Figure 1**); later you can do it more relaxed in the saddle. Also say, "Whoa," and remember that this is the only time to use that word because you want to emphasize that it's only for stopping.

There should be no tension on the reins (**Figure 2**) when you say, "Whoa." Hold a rein loosely in each hand with the slack ends crossed in front of the saddle so they lay on each side of the horse's neck. That way you can reach down and lift more rein up quickly when you need to. If the horse doesn't stop, give it a correction: Use both your hands to give one or two quick tugs on the reins at the same time. These "pops" don't hurt the horse, but the horse will notice. The horse won't feel the correction if you are dragging or pulling on the reins all the time. Riding with light hands and a quick correction work the best.

It's okay to give the cues slower in the beginning so there is good communication with the horse, but be sure to have the horse's attention before cueing it to stop. Always get its ear before asking the horse to do anything, and don't catch it off guard.

To correct a horse that just slows to halt instead of stopping, do the same thing: Drive the horse forward at the walk and say, "Whoa," while simultaneously leaning back in the saddle. If the horse doesn't stop immediately, give it a couple of quick pops or jerks with both hands to correct it so the horse stops.

Repeat the lesson several times. Ride forward again, say, "Whoa," and put on the brakes by pushing your feet forward while leaning back. If there is no response, give a couple of quick pops on the reins that are sharp enough for the horse to feel them.

The correction is all in the wrists, not the arms. Don't yank back on the reins, just flex your wrists down (**Figure 3**) in a quick movement. Your hands should be low, and try to keep them stable and unchanging in one spot. Definitely don't pull back with your arms. (**Figure 4**). Let the wrist action do the popping. The correction isn't meant

Figure 4. The rider is pulling with her arms, which is more like restraining the horse instead of correctly popping the reins. She also is not cuing him to stop.

Figure 5. The rider leans back with her feet pushed forward—the cue to stop the horse—and flexes her wrists to correct him because he didn't stop. This is the right way.

Figure 6. You can give a grain reward to the horse while mounted.

Figure 7. The horse brings a foot forward so he can balance after the stop.

Figure 8. The rider does the "Whoa" without using the reins.

STEP-BY-STEP

Figure 9. Practice stopping by a different panel of the roundpen each time.

Figure 10. Back up a horse to correct it if the horse didn't stop perfectly.

Figure 11. The horse is starting to bend at the poll and stopping quicker when the rider cues him to stop. She still has to pop the reins because it isn't quick enough.

to be punishment and doesn't hurt the horse because of the snaffle bit. It is only one or two quick pops or sharp bumps with both hands on the reins, and it has to be done immediately. By quickly flexing your wrists down a couple of times, the horse will notice it and want to avoid it, but it isn't painful like a hard yank on the mouth would be. It's just a simple correction with a light snaffle bit.

There is another type of correction I was taught as a kid, which is pulling back hard on the reins and sawing on them back and forth. This is much too harsh, and the horse usually throws its head up to brace against it. With my method, the horse will learn to stop and even back up nicely by just lifting the reins. The horse will do it if it understands what is wanted and when to do it.

To continue the lesson, ride forward again, say, "Whoa," and lean back while bracing against the stirrups. If the horse responds by slowing down instead of actually stopping, pop once on the reins.

Walk the horse around a little before asking it to stop again. When you are ready, lean back and say, "Whoa." If the horse doesn't stop right away, pop the reins once (**Figure 5**). You can pat its neck afterwards for trying. What you don't want to do is overcorrect and get the horse to toss its head. It also isn't any good to do the correction too late; it must be done with the right timing. Make the correction and then relax.

Continue practicing until you get a good stop. Rein the horse around in a big circle and have the reins in your hands ready for a correction. Then push your feet forward, lean back, and say, "Whoa." When the horse stops right away on a loose rein, that is complete control, so there is no correction and you can give it a handful of grain as a nice reward (**Figure 6**). Only reward the horse when it stops really well and with perfect timing. If you give a reward when the horse does it just so-so, then it will end up doing it so-so all the time.

One more time, walk the horse around and when you are ready, say, "Whoa," and lean back. If the horse slows before it stops now, back it up a couple of steps. There's a big difference between a slow stop and a quick stop. Don't reward a slow stop with grain. Remember: with this technique, the horse must stop perfectly for a treat.

Concentrate on your timing for giving cues. It's best not to stop the horse too quickly or when it's in midstride, which can set the horse up to make a mistake. Don't correct the horse if you didn't cue it correctly, but don't reward the horse either.

We want an immediate stop. If the horse takes another couple of steps after you correctly cue it to stop, you need to catch that and immediately correct the horse before it takes another step. By allowing the horse to take a step or two, it is going to be walking off on you. If the horse happens to be in motion and can't stop without putting another foot forward, that's okay because it will need to complete the stride to maintain balance. If the horse puts one foot down or holds it forward at the "Whoa," and then gradually brings the other foot up to balance (**Figure 7**), that's okay too.

The horse's timing should be spot-on. It's not okay if the horse tries to beat the correction by slowing down or stopping before the cue to stop. If the horse is corrected at the right time, it will learn that the best way to beat a correction is to stop right when you ask it to. When the horse does it correctly and stops on cue, release the pressure right away and give it a reward.

Ride for a while before you lean back again and say, "Whoa." If the horse stops perfectly and right away, give it a handful of grain. I like to let my horse stop and think about it when he does it right and enjoy his reward before riding him around again.

Before moving on to a trot or a lope, the horse should have progressed to stopping the second you lean back and say, "Whoa," and it should stop well enough that it doesn't have to be corrected. Make sure your horse is dependable with this before moving on to the next phase.

When you think your horse understands the "Whoa" cues, try something else. As the horse walks, drop the reins entirely, and give the cues: Lean back in the saddle, brace your feet forward against the stirrups, and say, "Whoa." (**Figure 8**) If the horse stops right away, it means you are getting more control, so reward your horse with a treat. You want it really ingrained in the horse's mind to do it perfectly, so be consistent and reward the horse when it does it right.

When riding your horse around, be careful not to choose the same spot where you stopped it before. The horse will get used to that and anticipate it. Then you're not really teaching the horse to respond to the cues. It's easy to know it's anticipating it because the horse will start to slow down, but keep the horse at a consistent walk. When doing this in the roundpen, just pick a different panel to stop the horse each time (**Figure 9**).

The biggest problem I have teaching this lesson is when owners get ahead of their horse and think, "My horse is doing this okay, but I really want to do it at the lope, so I'll go ahead and lope my horse, then make him stop." Be careful not to do that, and don't think two or three perfect stops will be enough to try

You should be an active rider, telling the horse what to do and then following through with it.

a sliding stop on a lope. It just isn't going to happen. We want to be consistent before going to a faster gait. Take the baby steps first and don't move on too quickly.

When your horse becomes consistent and you believe it is ready to move on, ride the horse around at a quicker walk and cue it to stop. If the horse stops but not right away, don't give it a treat. Lift the reins and back up a few steps (**Figure 10**), then rein the horse around the other way and try again.

When your horse stops correctly at a quick walk, then you can bring it into a light jog trot and tell it "Whoa." When the horse can do it needing only a light correction, give it grain and a pat on the neck. Practice until you get a perfect stop by just saying, "Whoa," and pushing your feet forward. It gets easier to keep the reins loose and trust the horse to stop when you get a response like that. When the horse does it with the reins loose, you have complete control.

The nice thing is that, with consistency, the horse will eventually bend its head at the poll when it stops (**Figure 11**). We don't want our horses getting peg legged (stiff in the front) and throwing their heads up, so that's why we start with baby steps.

As you go on with this, you can gradually phase out the treats, and the horse should continue to stop when you ask it to. Treats are just to help with training and conditioning to do the right thing.

When your horse is ready to stop without the verbiage, give the motion cues without saying "Whoa" and see if the horse can do it perfectly. When it does, it's obvious you are communicating well. Then bring the horse back to the walk again and say, "Whoa," without picking up the reins. If the horse stops well, that is positive and makes this a good time to quit.

SUMMARY

Training for stopping should begin in the roundpen; later, it's okay to go to an outside arena to practice. When your horse does a perfect stop without you needing to say, "Whoa"—only bracing your legs and leaning back—then you can go to busier surroundings. You will find it's a bit different at each place, and there will be different reactions from your horse that you will need to work on. The roundpen is more restricted, the outdoor arena is a little bigger, and then going on a trail ride is completely different. It's fun to work on training during a trail ride as opposed to just being a passenger. You should be an active rider, telling the horse what to do and then following through with it.

Take your time and make sure your horse is paying attention. It's not fair to ask the horse to stop when it is looking at other horses or attentive to something else. Make sure that the horse stops well and that it does it more than a few times before trying it at a trot or lope. Eventually the horse will get conditioned to stop, no matter what the circumstances.

I don't show my horses, but I do train horses to get their rears under them so they can do a sliding stop. When I do that, I ride them down a gradual hill and get them into a trot or lope so they learn to get their rears under them, but I always start with baby steps.

I especially enjoy this time with my horse when it's a beautiful day and I have time to train. It's great when the horse is paying attention, even if the horse doesn't know a lot. If you have a horse that catches on quickly and you have its attention, the time with the horse can be great. Enjoy it, and your horse will love it too.

Be safe and have fun!

Total time for this session: 9 minutes

Dennis directed these exercises:

1. The horse was taught to stop by using a verbal command and the rider leaning back in the saddle.
2. The horse stopped without the verbal command and also without the rider touching the reins.
3. The horse stopped from a slow trot by the rider just leaning back in the saddle.

CHAPTER 3

PROBLEM SOLVING

The goal of my problem-solving training methods is to have a steady, cool, and calm horse that's trusting and trustworthy. I designed these methods with several achievements in mind: stay safe, have fun, and train your horse quickly and efficiently. I'm going to show you a variety of issues and how to fix them. Keep in mind that common sense is crucial to problem solving.

Before we get any further, however, I want to impress upon you the importance of establishing leadership in the roundpen first before going on to any of the lessons in this chapter. Leadership is a form of communication needed before we can resolve any problems with your horse. If we have no means of communication, we can't go any further. It wouldn't be fair to the horse or us. We would get frustrated and turn to breaking our horse instead of training it. So take your time and start in the controlled, safe environment of the roundpen. If we reward properly and quit on a good note, our control

is going to be phenomenal. These details will create a bond with your horse that will never go away.

OVERVIEW

I use only a few tools for the problem-solving training sessions. They are simple, inexpensive, and can be purchased at any hardware store. The first tool is nylon string. I use it with a lot of techniques. The second is a metal snap, which I attach to the string for safety reasons in some lessons. A third helpful tool is actually my baseball cap. It's important for me to wear a cap instead of a cowboy hat.

Last but not least is the grain I use to reward the horse. The concept of treating our horses is different than what we were taught since we were kids, which was, "Don't give treats to horses or they will get aggressive." I practiced it for years and never gave any treats. Then I found out that a reward system cuts the training time immensely. In

My problem-solving techniques address catching, bridling, and mounting your horse.
Nick Vedros, MindFire Communications

DID YOU KNOW?

THE FLORIDA Cracker Horse was America's first cowhorse, working cattle in the harsh subtropic environment of Florida since colonial times. Though it is now critically endangered, ranchers still place great importance on the work of these enduring little horses. *Victoria Tollman, Equus Survival Trust*

some circumstances, like despooking, you can get results that you never could accomplish without rewarding the horse with treats. The clearest language you can speak with your horse is the reward system.

If we continue on this path of training rather than breaking horses, we will have a happy and healthy relationship with a trusting and trustworthy partner: our horse. Stay safe and have fun!

STAND STILL WHILE MOUNTING

A well-trained horse stands calmly, stays in one spot while you are mounting, and stays there until you are ready to ride. The horse should not move at all until you are settled in the saddle and have cued it to go. There should be no concern about the horse walking off until you want it to.

If a horse does move while it is being mounted, it usually continues to move; some horses take advantage of that and just keep walking before you can get on. When it does this or if it turns its head toward you, swinging its rear away while you are mounting, it makes it difficult to get in the saddle. This can be quite irritating, and it can also be dangerous. Even if the horse just takes a few steps forward when you try to get on, you can be left hopping with one foot on the ground and the other in the stirrup (**Figure 1**). The horse could run off with you dangling there and if your foot gets caught in the stirrup, you could get dragged. Suddenly you are facing a life-and-death situation. So take the time to train your horse to stand still properly while mounting. Remember: it doesn't affect just you; it affects everyone who rides your horse.

For this lesson to work, your horse needs to know how to take grain as a treat from your hand politely because we will be using a reward system. I discussed treating on page 59 in the Head Down lesson, but here's a refresher: When you offer grain to a horse, watch to see if it takes it nicely. If it doesn't and gets too aggressive, pop the horse with your closed hand enough for the horse to lift its head (**Figure 2**). I do that any time a horse is pushy with treats. I don't do it hard enough to cause bleeding or damage, just enough to get the horse's attention and let it know I don't like the way the horse took the treat. Then I tell the horse to relax and offer the treat again until the horse takes it gently (**Figure 3**). I'll be polite as long as the horse is.

We want to teach the horse the cue to assume a stance that is safe for mounting and reward the horse when it does it correctly. For that reason, keep some grain conveniently located in a saddlebag on both sides of the horse's saddle.

THE LESSON

Start by standing next to the horse, facing the saddle with the left rein in your left hand and hold it against the horse's shoulder. Place your right hand on the saddle horn. Push on the horse's shoulder with your left hand and rock the saddle rapidly back and forth with your right hand at the same time. This is the cue to stand still.

With the pressure on the horse's shoulder and the shaking of the saddle, the horse has to brace itself and spread its legs apart enough that its weight is placed evenly on all fours. This is called squaring up, and this is what allows a rider to mount safely.

If the horse moves its feet only a little, but not enough to stand squarely (**Figure 4**),

You can put on your coat, get a drink, or whatever, but until you're ready, don't let the horse move anywhere: Make it stand there.

do it again. If the horse moves to get all four feet directly beneath it with its weight evenly distributed on all fours, give the horse a bite of grain from the saddle bag to let it know the horse did what you wanted, and then mount up.

The pushing and shaking is a lot, not just a little—enough to make the horse move and stand straight. If the horse's legs are crossed (**Figure 5**) or too close together when the rider mounts and the rider doesn't notice and correct it, the horse is going to walk off. That would be the rider's fault. It's just like a person who is standing with more weight on one foot than the other or with one foot ahead of the other: If the person is pushed, that person is going to fall, so he or she has to move and catch himself or herself to stay standing. It's the same with a horse. Shaking the saddle back and forth and pushing on the horse will cause the horse to square up its feet. The horse will change its relaxed stance to a four-squared one.

After the horse has squared up successfully and you've mounted it, walk the horse around a little before trying it again. When you do dismount and try again, be sure to get close to your horse for your own safety. Push the saddle horn back and forth, away and then toward yourself, while pushing with the other

hand holding the rein on its shoulder. Don't quit pushing until the horse stops moving and spreads its legs to accommodate the movement. That's when to quit, let the horse relax, and mount up (**Figure 6**). You can either treat the horse from the ground before you mount, or you can treat the horse from the saddle once you are mounted.

I square up a horse for two reasons. First, if the horse's feet are too close together, it is going to move to compensate for the added weight pulling on its side when the horse is mounted, so the horse needs to stand squarely. I want to check where its feet are at, which is important to know before I mount. Shaking the saddle confirms the horse has a solid stance and is ready for me.

The second reason to square up a horse is simply to cue it to stand still until you get on. Always shake the saddle horn, no matter what, even if the horse is standing squarely. That's the horse's cue so it will know to stand still while you get on. If it looks like your horse is already standing squarely, shake the saddle anyway. It's a typical scenario when the horse looks like it is already squared up and it doesn't seem like there are any issues with mounting, but never assume your horse is ready. Always do the cue for the horse to stand still.

Only when you are mounted and ready should the horse be allowed to move. You can put on your coat, get a drink, or whatever, but until you're ready, don't let the horse move anywhere: Make it stand there.

After the horse is rewarded, dismount and then lead it off a little and do it again. Shake the saddle until the horse moves its legs apart. Mount up and reward the horse with a handful of grain. If the horse takes a step or two after you are mounted because a foot was not

QUARTER HORSES and JROTC cadets at San Marcos Academy in Texas recreate an authentic 1870s-era U.S. cavalry. As the Harper J. Moss Memorial Mounted Color Guard, they present colors during the Annual Federal Inspection of the Corps of Cadets.
Chris Hurd

STEP-BY-STEP

Figure 1. The horse walks away when the owner tries to mount.

Figure 2. The horse is annoyed and lifts its head after I popped his mouth with a closed hand for being too pushy with treats.

Figure 3. The horse asks politely and has no trouble taking the treat after being corrected.

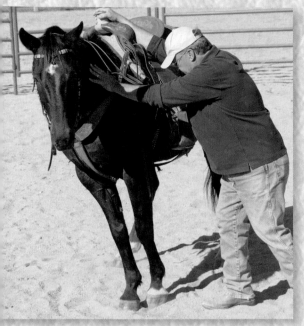

Figure 4. I shake the saddle and push on the horse's shoulder until the horse moves to brace against it. Although the horse's rear legs are well apart, its front legs are too close together.

Figure 5. The horse stands with its legs crossed, which will cause the horse to move when the rider gets on.

Figure 6. The horse stands with its legs apart now, so I help by treating the horse while the rider mounts.

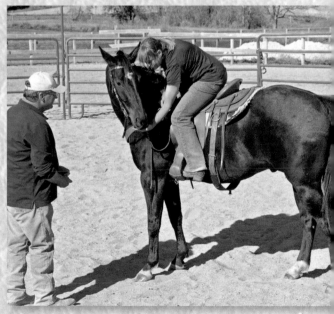

Figure 7. The horse didn't move, so the owner treats the horse from the saddle.

squared up perfectly, then dismount to repeat the process and wait a bit after you mount to make sure the horse doesn't move before giving it the treat. If the horse places its front legs wide apart and takes just a little step after you mount to adjust the stance but doesn't walk off, I would still give the horse a treat.

Keep practicing until your horse gets it. Remember to push hard with your left hand against the horse's shoulder and your right hand on the saddle horn back and forth. When the horse is squared, mount up. If the horse doesn't move once you've mounted, bend down from the saddle to treat it (**Figure 8**), which can be done from either side of the horse. Ride the horse around, dismount, and do it again.

The steps are:

1. Stand close to your horse and look at all four of its feet to check where they are.

2. Get your left hand (holding the left rein) on the horse's shoulder and your right hand on the horn.

3. Push on the horse's shoulder and shake the saddle, rocking them forcibly. This tells the horse to get ready, set itself up, and stand still.

4. If the horse squares up its feet, give it a treat right away. You can quit treating from the ground when the horse is consistent; then just mount and treat from the saddle instead.

5. Ride the horse around, then dismount to do it again.

SUMMARY

Practicing this exercise gets a horse to really square up, which will accommodate even a heavy person getting on. If you use the reward system properly, the horse will have a positive response and continue to stand still while being mounted even after the treats are gradually phased out.

With this method, we're controlling movement and establishing leadership in a way that makes sense so our horses can understand there is no pain involved and they will be rewarded. It's pleasant for them to learn what we want.

Before you mount, make sure your horse is squared up, give it a treat in the correct manner and timing, and always quit on a good note.

Total time for this session: 10 minutes

Dennis directed these exercises:

1. The horse was taught to stand squarely on cue and not move during mounting and after it was mounted.

2. The horse was corrected when it didn't square itself up enough.

3. The rider was taught how to cue the horse to stand still.

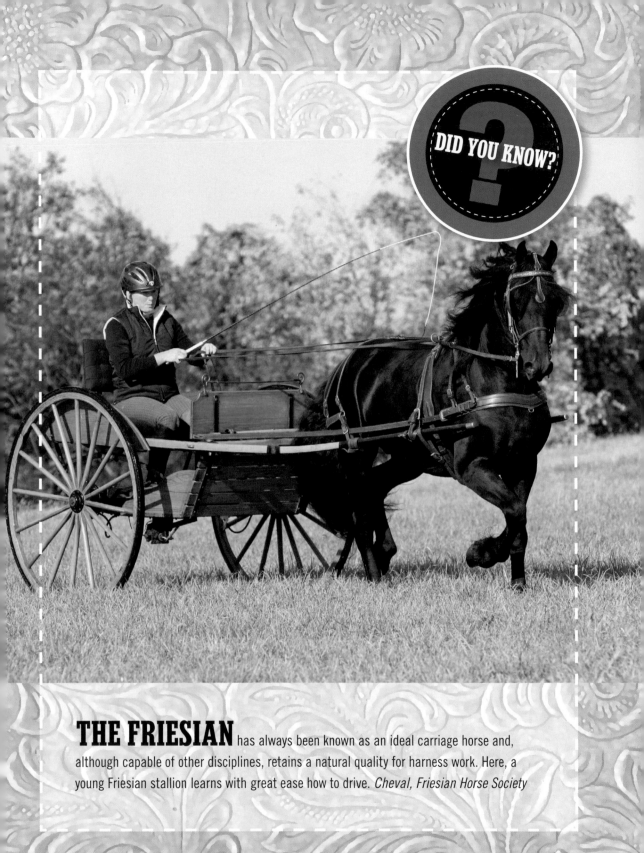

THE FRIESIAN has always been known as an ideal carriage horse and, although capable of other disciplines, retains a natural quality for harness work. Here, a young Friesian stallion learns with great ease how to drive. *Cheval, Friesian Horse Society*

STEP-BY-STEP

Figure 1. I position myself next to the horse with the bridle and grain.

Figure 2. My right hand holds the bridle between the mare's ears to keep her head down. I bring the bit up to her mouth along with grain in my left hand, which I give to her.

Figure 3. The horse lowers her head and turns away, resisting the bit.

ACCEPTING THE BIT

A problem that some horses have is accepting the bit when being bridled. Teaching this to a resistant horse that raises its head out of reach is kind of tricky and takes coordination. When we have a horse that is hard to catch, our first instinct is to go after it and grab it. It's the same thing here: When we have a horse that is fighting to get its head out of reach, our instinct is to fight harder to get its head down. The horse is thinking, "You can fight all you want, but I'm not going to do it!" We'll soon learn we can't force the horse to do something it really doesn't want to do because the horse is so much stronger. When we can resist our impulse and work calmly and patiently, we are on the path to teaching the horse how to accept the bit.

Today, I am in the roundpen with a mare whose owner has difficulty bridling her. She can't get the horse's mouth open to accept the bit, and the horse makes it worse by lifting her head and turning it away.

"When I put the bridle on, she pushes her nose up in the air. It's a two-man job just to get the bridle on!" the owner complains. "The more I struggle with her, the higher she lifts her head."

I understand. I have struggled like that before trying to bridle a horse that didn't want to be bridled. The horse smacked me in the teeth with its head, and that's not fun.

Before working with a horse on this issue, the horse should be checked by a veterinarian for any dental problems that could be aggravated by the bit. If your horse consistently refuses the bit, she could have teeth problems that are causing pain.

If you have done roundpenning correctly, your horse shouldn't need to be haltered; otherwise, she may have to wear a halter with a leadrope for this exercise. This lesson will work much better if your horse has been taught the cue to bring her head down. (See Head Down lesson, pages 59–65.)

THE LESSON

This lesson begins by standing next to the horse with a bucket of grain on the ground to your left. Take a handful of grain in your left hand and hold the bridle in your right hand. (**Figure 1**). Lift the bridle over the horse's neck, bring it down past her right ear, slide it forward to place the bit under her chin, and rest your arm between the horse's ears and hold the bridle. (**Figure 2**).

If the horse tries to lift her head, the weight of your arm resting between her ears will apply pressure on her poll to keep it down. Though your horse will find it difficult to lift her head, she might still move it around to avoid the bit, even with the weight of your arm there. Don't fight her, but don't release the pressure on her poll either. Entice her to move her head down with the grain in your hand, all the while saying, "Head down."

The horse will have to lower her head to take the grain. Try to get your thumb and finger in the corners of the horse's mouth to open it. If she does, give her the grain and guide the bit in. If you are able to get the bit in, don't finish bringing the bridle over the horse's ears. Just hold it there. You will need to take the bit out and put it back in repeatedly, so keep your right hand with the bridle crown between her ears. To get the bit out, say, "Bit, bit," while feeding the horse the grain. If she takes the grain, the bit will drop from her mouth. Remove it and give her more grain as a reward. Try to put the bit in her mouth again using another handful of grain to entice

Figure 4. My left hand is on the left side of the horse's mouth to keep her head positioned there.

Figure 5. I offer her grain, and as she reaches for it, I slide the bit out.

Figure 6. With my right arm over her head on the poll and my fingers pressing down in the corners of her mouth, I attempt to slide the bit in again.

Figure 7. My thumb and fingers press down, applying pressure on her gums to open her mouth.

Figure 8. The horse keeps turning her head away. I won't correct her when she tries to avoid the bit.

Figure 9. My hand holds the bridle between the horse's ears while I try to keep my head away from the horse's head as I give her the grain.

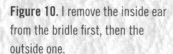
Figure 10. I remove the inside ear from the bridle first, then the outside one.

her. If she moves her head away, put your right arm on the other side of her head and pull it back in to you.

Some horses resist by lowering their heads closer to the ground (**Figure 3**). This is a dangerous position. If you are not careful and bend with your head over the horse's head, you could get whacked in the head. I did that once and it was a big mistake. I had to get new teeth afterwards. So keep to the side as much as you can with your left hand on the left side of the horse's mouth to keep her head positioned there. (**Figure 4**)

If the horse lifts her nose up high, don't fight her. Keep repeating, "Bit, bit," and try to get your thumb and finger in her mouth to apply pressure down. If you can do that,

you should be able to get the bit in. Hold the crown of the bridle in your right hand between the horse's ears, but just hold it there to keep the bit in place and reward her with grain from the other hand.

Then say, "Head down," and offer her grain again. If she lowers her head to get the grain, say, "Bit," and remove it. The horse should drop it easily out of her mouth when she reaches for the grain or chews on it (**Figure 5**).

Be careful not to bump her teeth with the bit when you are removing it. Some horses resist because they don't like the taste of fingers in their mouths, which makes the process more difficult. However, horses will usually take the bit easily and release it when they learn they get grain along with the bit.

Figure 11. The mare's head is at a comfortable level, and she takes the bit easily now with no resistance.

This lesson requires repetition and verbal commands. I like to use verbal cues a lot to help with the training process. Horses learn quickly what to expect when we can tell them, especially when the verbal cues are combined with repetitive motions.

Put the bit in again by placing your right hand, which is holding the bridle over the horse's neck, and sliding it forward, bringing her head closer to you. Position your right arm over her head on the poll with the bridle crown in your right hand. Say, "Bit," and guide the bit to her mouth with the other hand. If the horse sticks her nose out to the side, trying to avoid the bit, hold it by her mouth with your thumb and finger in the corners of her mouth (**Figure 6 and 7**) and apply pressure.

Never fight the horse. Just try to keep her head between your right and left hands with the left hand holding the bit. If the horse lowers her head to the ground again, click to her to bring it up, and then try again.

This is the process:

1. Hold the bridle between her ears with your right arm resting on her poll.

2. Say, "Bit," to the horse.

3. Guide the bit to her mouth, holding both the grain and the bit in the left hand.

4. Insert your left thumb and finger in the corners of her mouth to apply pressure and give a little grain at the same time.

5. Guide the bit into her mouth and give a reward.

6. Say, "Bit," again and remove the bit by offering her more grain.

Your horse should start to accept the bit without a fuss if you give her a bit of grain right away as a reward. Continue putting the bit in her mouth and taking it out without completing the bridling, just repeating the bitting process.

Don't correct the horse when she tries to avoid the bit. With this training technique, you want everything to be positive (**Figure 8**) when you present it to her.

By now, the horse should know you have the grain and be more receptive, although you may have to keep bringing her head back to you with your right arm extended over and around the other side of her head if she turns it away. Say, "Bit," as you bring it to her mouth. Have the grain and bit cradled in your left hand ready when she allows you to get your thumb and finger into the corners of her mouth. Slide the bit in and give her the grain at the same time. Then say, "Bit," take the bit out of her mouth, and give her the rest of the grain. Keep the bridle in your right hand with your right arm resting between her ears so she doesn't jerk up her head. Remember, never put your head over the horse's head (**Figure 9**).

Keep practicing and repeating. Never rush through the steps of a lesson. If your horse continues to move away from the bridle, don't fight her because that would be a negative experience. Continue to say, "Bit," as you get your thumb and finger in the corners of her mouth. It may be a taste factor that makes the horse fight any fingers in her mouth, but try to keep her head turned toward you at a comfortable level.

After you've successfully inserted and removed the bit enough times for the horse

If you're in a hurry to do anything— load, bridle, saddle, or anything else with your horse—don't do it. It will turn out to be a negative for everybody.

to accept it better, you can progress to putting the rest of the bridle on. First, place the horse's off ear in, which gives more control of her head. If she resists and raises her head, say, "Head down," (with your right arm still between her ears) and offer the horse grain at a lower level. Be careful to stay safe and give it to her without bending over her head. Then get the horse's other ear in the bridle. If your horse still insists on bringing her head up so high that it makes it tough for bridling, applying pressure to the poll will help to get her head down. If you touch the poll with your fingers, she should drop her head more. Eventually she will relinquish and take the bit with no struggling, and you can put the entire bridle on her successfully.

Then give her a break. Once the bridle is on, walk your horse around a little so she doesn't get bored or cranky and to help get her mind off the lesson. When you come back to practice after a minute, she will be more refreshed.

To remove the bridle, start from the top of her head, for safety reasons, and remove her inside ear from the bridle first (**Figure 10**). Then remove the outside ear. Give her grain and say, "Bit," as you gently slide off the bridle. If she is reaching for the grain, the bit will drop out easily. Hold the bridle in your right hand on the opposite side of her head with your right arm resting just behind her ears to keep her head at a comfortable level. Then reward her with grain again and allow her to chew on it for a minute and think about it.

It means so much to give the horse a chance to think about what you just did, which brings a positive aspect to working on the issue. One thing I have to say, if you're in a hurry to do anything—load, bridle, saddle, or anything else with your horse—don't do it. It will turn out to be a negative for everybody.

By this time, your horse might be relishing the grain handouts; if she gets pushy when you reach for more grain or crowds, push her head away. This is the only correction I advise for this particular lesson. Don't give her any grain until she earns it.

If you keep practicing, your horse will start turning her head toward you and reaching for the grain in your hand. Eventually she will lower her head to your level, allow you to put a thumb and finger in the corners of her mouth, and accept the bit while taking the grain at the same time. You will be able to put the bit in with no struggle (**Figure 11**).

It should also get easier to take the bit out of her mouth. Say, "Off," as you remove the bridle from her ears. Offer her grain and say, "Bit," to take the bit out. If she is chewing on grain, she will allow the bit to slide out of her mouth, so reward her again.

People ask, "Do I have to give treats every time I bridle my horse?" Until you can bridle your horse easily, yes, you do; after that, only do it every fourth, fifth, or tenth time, or as many times as it takes for the horse to learn it. Rewarding behavior that a horse just learned reinforces the positives of the lesson. They don't really forget, but they might try to cheat

and get a treat without earning it. Be consistent and reward only when it's earned.

SUMMARY

As a kid, I was taught to show the horse who is the boss and I still do that, but in a positive manner. When I grew up, showing who was the boss meant pain and intimidation for the animal. There's no reason to do that. When the horse starts fighting, just relax, let it fight, and before long it's going to realize that struggling is useless.

After working with the mare I trained today to get her head down for bridling, I encouraged the owner to teach her horse to get the horse's head down by using the method with the leadrope and pressure on its poll (see Head Down lesson, pages 59–65), then work on the bitting issue.

The owner is happy with her horse's progress: "I think if I keep working with her, she'll get even better. She's keeping her head down so much more now than before, when she had her head too high to reach. Not only her nose, but her head was too high, so dropping her head is really going to help."

Remember to reward using the treating system and stop when the horse gives a positive response and relinquishes. Always quit with a positive finish and don't overwork the horse. That way there is no resistance when you bring her back to practice again.

| Total time for this session: 10 minutes |

Dennis directed these exercises:
1. Taught the horse to lower her head for bridling.
2. Taught the horse to accept the bit without raising her head, lowering her head, or otherwise resisting.
3. Taught the horse to give up the bit easily.

DID YOU KNOW?

THE ANNUAL Pony Swim from Assateague Island to Chincoteague Island off the coast of Virginia and Maryland is a popular event drawing people worldwide. These wild ponies are descended from herds that existed on Chincoteague more than four hundred years ago.
Carien Schippers, imagequine.com

HEAD TOSSING

Head tossing may seem like a minor problem, but it is a precursor to a much more dangerous issue. Once a horse starts tossing its head, it gets lighter in the front end, which means the front hooves start coming up higher and it could rear. The dangerous part, where a lot of people get hurt, is when the horse flips over backward on them.

A gelding was brought to me that had a problem with tossing his head up every time he was asked to go to a faster pace. I had worked with this horse previously in the roundpen, so I had already established leadership and got control of his movement. But I saw that every time he went from a trot to a lope, he brought his head up. This could develop into something worse unless we correct it immediately.

I use a simple technique to correct head tossing. It also fixes rearing, which is a dangerous issue. My technique causes the horse to correct itself, and there is no pain or harshness involved.

This lesson requires the use of a solid nylon string or cord that has been cut to about six yards long. You don't want it too short. Don't get the nylon that's wrapped in cotton because it will break. (**Figure 1**). Tie a snap or clasp to the center of the string with a simple loop knot and double over the string so both ends are even (**Figures 2, 3, 4, 5**).

THE LESSON

Begin the lesson in the roundpen and have the horse saddled and bridled. It's preferable to have the reins removed so they won't get in the way; otherwise, tie them up around the saddle horn. Take the snap that is tied to the string and clip it on the D-ring in the middle of the girth underneath the horse (**Figure 6**). Take one end of the string, bring it up between the horse's legs to the bridle, and run it through the ring on the side of the bit where the bridle attaches to the bit (**Figure 7**). Then run the string up alongside the horse's head to the top of his head behind the ears and drape it there for now. Do not tie the string to anything on the bridle.

Do the same with the other end of the string on the other side of the horse. Bring it up from the girth between the horse's front legs and run it straight to the other side of the horse's head, through the ring on the side of the bit (**Figure 8**), and up to the top of the horse's head behind his ears. Both lines now extend from between the horse's front legs (**Figure 9**) and run to each side of the bridle bit and up behind the ears. (**Figure 10**). Make sure both strings are the same length.

Tie the ends of the string (**Figure 11**) together in a simple bowknot behind the ears, allowing for enough length so there is adequate slack for the horse's head to be at a comfortable level.

The amount of string is important. It can't be too short or too long. You don't want to tie the string so short that the horse's head is cranked down; this would make him feel restricted. Then he might panic when he starts to move and do some potentially crazy things that are dangerous to him or you, like fighting it and flipping over backwards. Be observant and notice if the horse is standing relaxed with his head positioned a little low. Don't tie the string off with it at that level or it may be too short and too tight. The horse has to have enough latitude to extend his head at a normal level when he trots or lopes. He should only get corrected when he tries to bring his head up too high, so tie the string loose enough to allow some movement of his

Figure 1. This is the typical nylon string I use and the snap that attaches to it.

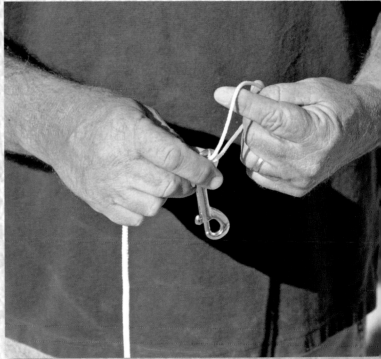

Figure 2. The snap is attached in the middle of the string using a simple loop in the string.

Figure 3. The string ends are passed through the loop until it fits snugly.

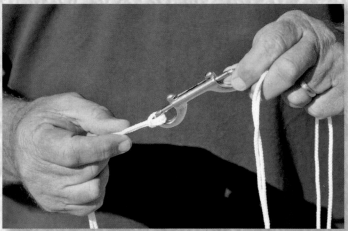

Figure 4. The string is properly attached to the snap now.

Figure 5. This is how the snap on the string should look.

Be on the lookout for the horse to start relaxing, which will be evident when the string under his neck goes slack. This means he is carrying his head lower.

head. You might need help from someone else to lift the horse's head up a little above the withers or whatever is normal for your horse and your type of riding.

The string also can't be too long, or he won't feel its pressure. When you tie the ends behind the horse's ears, make sure it is not on top of the bridle because that's where the correction will be and the horse needs to feel it.

When the horse tosses his head with the string in place like this, he is going to correct himself both at the bit and the poll. The horse won't relate it to punishment and it won't be harsh, painful, or intimidating. He is going to educate himself, and it will have no relevance to you whatsoever. What we are looking for is pressure and release. The horse will actually be pressuring himself when he lifts his head and the string bumps against his poll and mouth. He will correct himself when he feels that pressure and bring his head lower.

When the string is correctly in place, step to the center of the roundpen and kiss to the horse or motion with your hands to get him walking around. Give him a chance to respond and move, but if he doesn't, raise a whip to get him going. Start moving him around at the walk and let him toss his head

all he wants, which will make the string bump on the bit and on his poll.

For this type of roundpenning, I don't care which direction the horse turns. The thing I am concentrating on is the head tossing. Let the horse test the string as he moves around (**Figure 12**). We want him to bump it and then release the pressure, which is correcting himself. The horse will start doing this immediately, which he won't relate to you at all.

Be on the lookout for the horse to start relaxing, which will be evident when the string under his neck goes slack. This means he is carrying his head lower. When he does this, get him into a trot and if he bumps on the string again, notice when he starts releasing the pressure on it. The pressure is on both the horse's poll and mouth when he lifts his head. The horse will automatically correct himself by bringing his head down so he won't get bumped. It's obvious when he releases pressure because the string gets a soft dip in it.

Before moving the horse into a faster pace, make sure he is comfortable with it at the trot. He may still try to raise his head several more times and get bumped with the bit on his mouth and behind his ears as he trots. Be patient and wait for that to quit (**Figure 13**).

Then move the horse into the lope and watch for the same thing. We want him to release the pressure after he has bumped himself, which he may do several times or a lot. Let the horse lope a couple of times around the pen and bump himself all he wants.

Bring him back to the trot and allow him to feel the bump when he makes that transition. Trot him once around the pen and bring him back into a lope. What you are looking for is smoothness in his transitions.

When I worked the gelding in the roundpen before, he brought his head up every

Figure 6. I attach the snap to the middle ring on the girth.

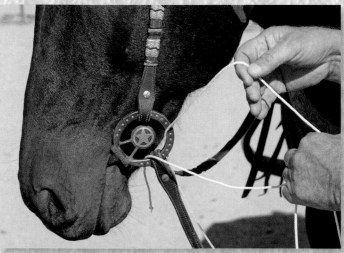

Figure 7. Run one of the string ends through the O ring at the side of the bit.

Figure 8. Do the same with the other string end on the other side of the horse's head.

Figure 9. Both ends of the string are now extending from the girth between the horse's front legs to the bridle.

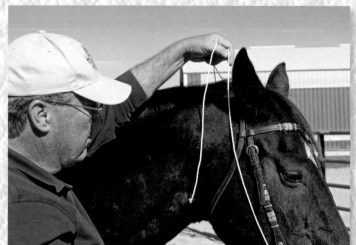

Figure 10. Bring the second string end also behind the horse's ears.

Figure 11. Tie the string ends in a simple bow knot behind the ears where they are not touching the bridle.

other time he went from a trot to a lope. We want our horses to make a nice, smooth transition from a trot to a lope without lifting and bumping their heads. Transitions between gaits should be easy, with no head tossing or raising of the heads. Horses that don't raise their heads during transitions are easier to ride, more controlled, and smoother. This is what show people like, and I like it when I go trail riding.

If your horse is bumping a lot, keep him at the lope until he releases the pressure on the string. If he continues trying to bring his head up, he will get bumped every time until he begins to understand. If the horse tries to stop and face you, start him trotting again. Soon he will carry his head low enough to avoid the bumping on his head and start doing the gait transitions better. Notice when he changes his gaits if there is any more head tossing or aggression.

We want the transitions from the walk, trot, and lope to be all the same. Bring the horse to a trot and then back again to a lope.

Practice with the horse until he does a nice, smooth transition where he doesn't raise his head and bump it. When he trots one circle in the roundpen without bumping his head, tell him to relax and then slow him down. We also want to see a nice headset at the walk. I don't show my horses, but I still like to see a good headset on them. High-headed horses are dangerous unless they're bred to be that way with different conformation.

Turn your horse around and move him in the other direction. If the horse trots several times around the pen with no bumping, get him into a lope. When there are only a couple of bumps and then he releases the pressure and brings his head lower, he is doing great (**Figure 14**).

The horse may still lift his head a little when he does a gait transition, but you don't want him bumping too heavily. If he does, tell him to relax, bring him to a walk, and start all over again. Move him from the walk to a trot and if he does a better transition, not bumping and keeping that release, that is great. Move him into the lope, and if he brings his head down automatically or only does a little bump, get him to trot again and see if it's any smoother.

If it's no smoother, make him turn and walk the other way. We want him to walk so he understands that you want a nice relaxed headset (**Figure 15**). Then ask for a trot, and if he does it with the string nice and loose or only puts a little bit of pressure on the string, lope him.

If he does it fine and his head position stays relaxed after one lope around the pen, let him stop and turn to face you.

To stop him, make a kissing sound and step into his path with your hand raised toward him like you did with the roundpen work. The horse should slow and key in on you, but don't look directly at his eyes if he gets close. The horse may then walk up to you (**Figure 16**) or allow you to approach him. If he is chewing or licking, you did well. Allow him to sniff your hand.

That's all you need to do for the first lesson on head tossing. If the horse is getting it, you don't want to make him run too much. Untie the bowknot on top of the horse's head, unsnap the string at the girth and take it off the horse.

SUMMARY

Keep the first lesson short. Don't worry about quitting too early on this one. What we want to accomplish is for the horse to learn not to toss or throw his head around. If he brought

Figure 12. The horse walks around the roundpen and starts bumping and testing the string right away. Notice the tautness in the string.

Figure 13. The horse lopes and bumps on the string.

Figure 14. The horse lopes with a better headset and tucks his nose in a little more now.

Figure 15. The horse walks with a nice headset with the string loose. He is also licking, which is always a good sign.

Figure 16. I raise a hand to invite the horse in and he comes to me while chewing.

his head lower, the horse should know that head tossing is a no-no, but he won't be completely fixed after just one lesson. I suggest training with the string at least one more time in the roundpen. That is our training school and where it would be best for the horse to start again next time.

Make sure when doing this the next time that the string is on the horse at approximately the same level it was today. Don't think that because the horse was great the first time, his head can be cranked way down closer to his chest; that's not fair to him. Start where the last lesson left off by allowing the same height for the horse's head as you had it today before tying him down any farther. It's best to train with this method by shortening the string inch by inch each time he does it right. Don't tie his head too low or he will fight it and may get hurt.

Keep the string in your pocket so it's handy whenever you need it. After a few lessons in the roundpen, you can use the string on the horse when going on a trail ride. This should be done several times until he becomes consistent. If the string gets tangled at all, you can always reach forward and untie it from the top of the horse's head. All it takes is one simple tug on the end of the bow knot to untie it. The string can also be used before a show or whenever it's needed to work on transitions.

If you quit the lesson on a positive note, the horse will follow you and stand quietly next to you.

Leadership—I love it!

Total time for this session: 8 minutes

Dennis directed these exercises:

1. Taught the horse not to toss his head during transitions.
2. Taught the horse to bring his head lower at all paces and in both directions.

HIGHLAND PONIES are known for their kindness and patience, as evidenced in this relaxing moment between mare and youth. Highlands are one of the nine Mountain and Moorland breeds native to Britain. *Victoria Tollman, Equus Survival Trust*

HARD TO CATCH

One of the biggest complaints I hear from owners is about horses that are hard to catch. I don't like the word *catch* because the horse should want to come to you. The problem is immaterial compared to the real issue: getting control. Like anything else we do with our horses, we have to gain control with them first, which means establishing leadership.

The main problem owners have is getting ahead of their horses. By that I mean they try to catch the horses quickly on their own timetable. I don't wear a watch when I go to get my horse because I'm going on the horse's time. If I catch a horse like it's a timed event, of course the horse is going to be more difficult to catch the next time.

Today I'm working with a mare that walks off when the owner tries to catch her. I am teaching her to let me approach through roundpenning. She is loose in the roundpen with no tack on because we don't want her to feel like she is in a trap. We want this to be a positive experience for her, so I have a bucket of grain handy in the middle of the pen and will reward her whenever I can.

THE LESSON

Take a small handful of grain and approach your horse in the roundpen while holding the halter and leadrope (**Figure 1**). Some people make the mistake of approaching their horses hiding the halter and leadrope behind their backs so their horses won't see them, but this can become a cue to your horse to run off. It's better to be upfront and honest with the horse. Approach her with the halter and leadrope in plain sight and tell her "Relax."

If the horse avoids you and walks away, move to the middle of the roundpen and drive

Like anything else we do with our horses, we have to gain control with them first, which means establishing leadership.

her off by swinging the halter and leadrope at her until she runs, then invite the horse in with a raised hand, which is inviting her to come to you.

If she doesn't run but turns her rear to you, it might be an indication someone tried to use grain before to catch her and she remembers. Wait a bit; when she turns her head toward you, approach again. If she turns her head away, it's her way of saying, "Stay away!" You can raise your hand toward her and kiss, asking her to move. If she stays there with her head turned away and won't budge, swish the leadrope at her until she moves away and then try to approach her again.

If she continues to move around the pen, tell her to relax again and take a couple of steps toward her, intercepting and blocking her.

If she stops and begins to give any signs of respect, such as licking, chewing, or pointing an ear at you, she may be more receptive. Kiss to her and if she turns her head to look at you, she is inviting you to come closer. Then it is okay to approach her, but stop if she turns her head away (**Figure 2**). Take a step at a time toward her and pay attention to her every move because you don't know when she's going to run off. Show her the halter and leadrope so she knows what you want.

Figure 1. I approach the mare with the halter and leadrope, but she runs away.

Figure 2. When a horse turns its head away, it means it doesn't want you any closer.

Figure 3. I only need to raise my arm a little with the halter and rope to cut the horse off from running one direction.

Figure 4. The horse has stopped and is looking at me, which is an invitation to approach.

Figure 5. I give the horse grain with the halter and leadrope in the other hand.

Figure 6. The mare allows me to touch her face.

If she runs off at any time, back off to the middle of the roundpen and let her run around it, but don't let her stop and ignore you. If she slows and looks at you, raise your arm off to the side to block her from turning. Some horses will key in on you when you raise your hand but then run off in the other direction if you take a step toward them. When a horse does that, raise your other hand to block her from running in that direction too. It will only take a few steps to do it, even if you are still in the middle of the roundpen. The more sensitive a horse is, the quicker the horse will respond to each little movement you make (**Figure 3**).

If your horse stops and turns her head toward you, try to approach and offer her a bit of grain from your hand. If she tries to move away when you get close, take a step in the same direction to block her and tell her "Relax." If she stops with her rear angled at you, only approach slowly and stop if she doesn't turn her head toward you. If she turns her head toward you when you stop but swings it away when you take another step, she is saying that when you stopped, that distance was okay, but when you took the next step, it wasn't.

Wait until she starts chewing or showing other signs of respect. When she does and turns her head to you, show her the leadrope. If she turns away, she is saying she doesn't want to be caught.

Essentially, you must be able to read your horse and what she is saying, then give her alternative choices if she won't let herself be caught. Here are some examples:

1. If she runs off when you get close, cut her off from running in that direction. If she turns the other way, block that direction too by stepping that way.

2. When she stops and looks at you, try to approach her (**Figure 4**).

3. If she stops with her rear closer to you than her head or moves in the other direction as you approach, shoo her away so she has to run. The last thing she is going to remember is being made to run because she walked off or was disrespectful. I don't like running a horse too hard, but the only alternative is to make her run if she wants to avoid you.

4. If she stops to face you, approach her with grain in your hand. Stop if she turns her head away. Look for other positive signs, such as licking, chewing, or putting her ears on you, before approaching again.

5. If she lets you approach and give her the grain but won't let you put the halter on or moves away, make her run around the pen a few more laps.

6. If she stands as still as a statue, not responding at all, drive her off and make her run. In all circumstances where she has to run, try not to let her run too much. Keep asking if you can approach by extending your hand toward her.

When the horse allows you to approach close enough, stroke her shoulder and offer her grain while holding the halter and leadrope in plain sight (**Figure 5**). Take your time and try to slowly put the leadrope around her neck. The biggest mistake you could make now would be to do any kind of discipline or punishment for running away. When you put the halter on, give the horse some grain from your hand, tell her to relax, and try to stroke her forehead.

You have accomplished a lot when the horse allows you to approach and catch her. That's when the leadership issue turns around in your favor.

The horse should only get a treat when you can touch her head, so if she moves it away, don't give the treat to her.

Bring her by the grain bucket and give her a grain reward from your hand. Then walk her around, back away from her, tell her to relax, and extend your hand to her with more grain. When she reaches for it, try to pet her head. She should only get a treat when you can touch her head, so if she moves away, don't give the treat to her. Try again to pet her head and if she lets you, give her the treat.

Take a step away, get more grain, and kiss to her. If she walks up to you without you pulling her in, show her the treat and try to touch her head again. Be careful not to move too quickly if your horse is sensitive about it or if she is still unsure about you. Keep trying until she allows you to touch her head, then give her another treat. In the same manner, touch all over her head, her poll, and her face (**Figure 6**) until she lets you without resisting. This is important: If the horse won't let you or anyone else touch her, she won't let herself be caught.

Lead her away from the bucket of grain and make her stop. Then step away and draw her back in to you with just your hand using the invisible rope (see page 16). If she is paying attention and allowing you to touch her anywhere on her face or neck without you reaching out to catch her, give her a treat. Even a head-shy horse will get more confident if she is rewarded each time she is touched.

Walk her around to let her relax and think about it. When you stop, she should stop at your shoulder. As with the roundpenning technique to establish leadership, she should be respecting your space instead of pulling away or smacking into you. If you click to her and turn to walk away, she should follow right at your shoulder (**Figure 7**) and keep her head within easy reach, even without any treats.

If she winces or draws back when you reach to touch her, remove your hand and don't give her any grain. Try opening your other hand and positioning it next to her head. If she turns her head into your hand and touches it, give her the grain. She is telling you that she is more comfortable coming to you than when you reach out to touch her.

If your horse is sensitive like that, bring your hand carefully up to her cheek or wherever she seems to be touchy, then draw it away. Don't let her take the grain without letting you touch her there. Continue bringing your hand up to and away from her face and when she allows you to touch her there, give her the grain. Don't give a treat without touching her. She should relax eventually and let you stroke her face or anywhere else.

This is a good time to practice other techniques for respect as well. Tell her "Head down" by directing with the leadrope under her head and giving her a treat lower to the ground so she will bring her head down to get it (**Figure 8**). (See Head Down lesson, pages 59–65.) The horse will be more relaxed when her head is down. Then walk her around.

You want this to be as positive an experience for your horse as possible. If you practice this gentle way of catching her, it will be easier to get closer to her each time.

STEP-BY-STEP

Figure 7. The mare follows respectfully at my shoulder without crowding or lagging behind me.

Figure 8. I work at getting the horse's head down by offering grain lower than her head.

Figure 9. I stroke the mare's face and prepare to put the halter on. She has a softer look in her eyes now.

Figure 10. I raise my hand to the horse as I approach, but stop when she turns her head away.

A horse like this won't be ready to be turned out in a big pasture after only one roundpen lesson, so practice in progressive increments. You could start next time by working her in a smaller area than the roundpen, then do it a few more times in the roundpen before working with her turned out in an arena, which is still an enclosed area. Only after you've mastered those steps can you move to turning her out in a pasture and catching her there.

Once you feel more confident in your horse, remove her halter and turn to walk away. If she follows you, step away from her and if she stays facing you, walk slowly back up to her, extend your hand again with the halter and rope, stroke her face, and wait till she chews and her eyes soften before getting the halter ready to put on (**Figure 9**). If she turns her head away, raise your hand toward

her cheek to draw her back in with the invisible rope. (See Roundpen lesson, page 16).

It can be frustrating if it looks like your horse is reverting back to where she was at the beginning, but horses have their own system of learning. They can learn quickly and progress from there, but then their learning comes back down. They can forget things or just refuse to do them. In that way, the learning curve of horses is up and down. Their learning is going to drop back down before it goes up again. Ninety percent of the time if you're doing the lesson correctly, the horse is not going to regress completely back to where she started from. Then the learning curve is going to come back up and continue going up from then on. Expect that horses are going to learn this way, so try to stay positive with them when it seems like they forgot everything they just learned.

Figure 11. The mare now has a relaxed head, soft eyes, and is chewing while I stroke her face.

At this point in the lesson, if your horse moves away or ignores you, drive her away, but just a little. If she is walking slowly, walk in the same direction because you want to draw her back in right away. Try to get close without pressuring her. If she stops, raise your hand to invite her in. When she turns to face you, step closer, but stop to release the pressure if she turns her head away (**Figure 10**). If you get her ear on you or other submissive signs, like licking or chewing, try to approach her again.

If she keeps her head turned away as you get close, try to stroke her shoulder, then put your hand under her head and turn the horse's head your way. Put the rope around her neck and put the halter on if she accepts it willingly. Tell her to relax and put your fingers on her cheek to draw her in with the invisible rope. If she turns to face you, turn around and lead her away.

Then go for a nice friendly walk. Practice stopping and don't let her crowd you or lag behind. Stroke her head and try to make this a positive experience for her. She deserves a reward only if she will allow you to touch her head. Then direct her head away with the leadrope, touch her on the cheek, and step back with your fingers pointing toward her cheek. Let her bring her head to your hand so she is touching you, and then give her the grain.

Just to make sure she is getting it, undo the halter again and hold it a minute. Then show the halter to her and if she seems to accept it, reward her positive response with a bit of grain. The idea is for her to associate the halter and leadrope with being treated kindly so she won't run away anymore. Another indication that she is relinquishing her opposition to you is if her eyes are softer, which means you are finally becoming the leader.

Practice drawing your other hand away from her face with the halter in your hand—which is inviting her in to you—and put the halter loosely around her neck. Then put the halter on if she accepts it, which she should be doing more easily now. Through control techniques like this, the problem of catching her will be alleviated.

SUMMARY

The horse's attitude should be different now than what it was in the beginning. A horse that runs away is usually hyper, carries her head high, and is anxious to get away. Your horse should now have a softer look in her eyes, be licking or chewing, and her head should be relaxed (**Figure 11**). You should be able to stroke her face enough to almost put her to sleep. It's fun training horses like this.

The issue of catching your horse is something you should continue working on until leadership is firmly established and you have total control. Rewarding her is important. After that, when you go out to catch her, if she starts walking off at all either when she sees the halter and leadrope or just to be by herself, then you should stop, click to her, and use your hand to draw her in. That's all it should take to get her to turn and face you.

People get frustrated when they can't catch their horses. They go out and try to rope them or get ten neighbors to try and help catch a horse. If they were to spend time just working with their horses instead, they would find out it's not only fun, it's rewarding, and the results are always the same.

Total time for this session: 18 minutes

Dennis directed these exercises:

1. Taught the horse not to run away from him.
2. Taught the horse to allow him to approach and halter her.
3. Taught the horse to willingly follow him as the leader.
4. Taught the horse to allow him to touch her head.

DID YOU KNOW?

RIDING IN a parade is fun when the horses aren't bothered by loud noises, crowds, and all the commotion that usually accompanies a celebration. The trust these beautiful horses have in their riders is expressed in their calm stance. *Unbridled Photography*

CHAPTER 4

DESPOOKING CLINIC

Despooking horses effectively requires patience and slow moves forward. By taking your time, you will gain more confidence, control, and bonding with your horse. There are no quick fixes, but you may be surprised at the things you will be able to expose your horse to without issue.

No matter what the issue, we always establish leadership first in the roundpen, which gives us control. Then we have a basis of understanding before moving on to anything else, particularly despooking. Once you are confident about the leadership issue and your horse is cooperating with you, only then you can move on to despooking.

OVERVIEW

Everything we do starts on the ground before we get on the horse's back. Always start despooking by touching the horse all over with your hand and remember to do proper treating.

A common tool I use for despooking is a training stick, which acts as an extension of my arm. I use a fiberglass one when I can, or even a whip will do. It's just to touch the horse all over, which will help to establish control and keep me safe.

The most important thing is not to get ahead of your horse; notice when the horse indicates it isn't ready to move on. Wait until its ears are focused on you so it is aware of when you are going to do something and what you are going to do.

If you are in a group, it will be fun to train because the horses help despook each other. If you are working alone, you have to pay extra attention to your horse: Don't go too far or too fast.

Despooking is an important part of keeping yourself and your horse safe.
Nick Vedros, MindFire Communications

A RIDER NEGOTIATES a fence during the cross-country phase on her Irish Sports Horse at the Withington Manor (UK) One Day Event. The Irish Draught is a primary component of the Irish Sport Horse. It is an endangered breed. *Karl Edmond*

Eventually you can expose your horse to a variety of unexpected objects and noises that will ultimately translate into things experienced on the trail. Milk jugs simulate the bump of saddlebags. Tarps prepare a horse for water crossings. You will go from touching with the hand to the banging of milk jugs to crossing over the tarp and on to spookier things like the sound of gunshots.

I reward with grain during despooking clinics. It's important to have the right timing when giving treats or the horse will get confused. Always reward as a positive aspect in training, to get something you want in return. Remember to quit on a good note, and stay safe and have fun.

PART I: HANDLING EARS, TOUCHING WITH BAGS, JUGS, AND OTHER SPOOKY THINGS

I'm in the barn with a group of owners and their horses conducting a despooking clinic (**Figure 1**). You may find a horse in the clinic that has the same sensitivity issue your horse has. Although it's fun to do as a group, you don't need to attend a clinic like this to get similar results if you use the same techniques.

Here is what we will be doing:

We begin by touching the horse everywhere on its head and body starting with our hands and then going to a training stick to touch its body. Then we will go to something simple—the saddle blanket—and do the same thing with it all over the horse's body.

We will desensitize step by step using progressively spookier items: plastic bags, milk jugs, and finally "the monster"—the tarp. With each step, the horse will get a treat when

it remains calm as it is being touched. We won't move on until the horse has accepted each object. All this is done while our horses are tied to the barn wall.

Once we get that part done, we will despook them outside to simulate things that could frighten them out on the trail. We will bring them through an obstacle course and desensitize them to loud noises.

All of these steps are done in one session of despooking.

Before we begin, I make sure each horse is haltered and tied high and fairly short to its own section of the barn wall. By that I mean that the horse's head is tied about three feet from the wall and high enough that it is level with their withers (**Figure 2**). The leadrope can't be too long or the horse would be able to move around and possibly hurt someone as we go through the desensitizing procedure.

If an adult owner has a child with them, the child can help by being the "treat master" (**Figure 3**). It's great to involve kids when training horses as long as an adult is supervising. Most kids want to help and can benefit by the exposure to training as well as the owner and horse. They can hold the grain bucket and treat the horse when the adult says, "Treat." Adults are then free to work on desensitizing the horse and don't have to stop and find the bucket of grain. The child should hold his or her hand down and closed until the adult says, "Treat." Then the child brings the treat up to the horse so it doesn't pull against its leadrope, opens his or her hand, and gives the treat.

If you don't have a little treat master, have your bucket of grain as far from the horse as you can so the horse isn't pulling to get it. If the horse pulls too much or is pushy when you give treats, pop up with your hand

STEP-BY-STEP

Figure 1. Owners in the clinic have their horses spaced apart and tied along the barn wall.

Figure 2. I check to make sure a horse is tied high and short enough to the wall.

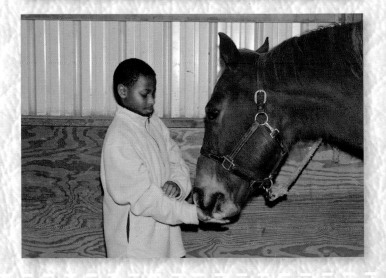

Figure 3. A child treats a horse while the owner works at desensitizing it.

Figure 4. I coach an owner on how to work her way up the horse's face to touch its ear.

Figure 5. An owner runs her hand over the forequarters of her horse while her other hand is braced on its shoulder.

Never step between the horse and the wall.

on its mouth so it is careful to take it politely. Make it ask for the grain, not demand it. If you pop your horse on the mouth, that is sensitizing its mouth. We want to desensitize our horses at the clinic, so stroke its mouth gently afterward and give the treat if it takes it carefully. Only give about a tablespoon of grain at a time.

THE LESSON

Starting together, the owners begin by touching their horse's forehead, poll, ears, lips, and around the eyes with their hands. The owner gives the horse a treat when it stands quietly for this. Always start on the head of the horse.

If your horse is skittish when you touch its face, give it a treat at the same time you approach to touch it. If there is an ear problem, back off and touch the horse on the head as close to its ear as it will let you and give it grain at the same time. This is better than rushing and getting ahead of the horse. For instance, if the horse is okay when you touch its nose, you might think it will be the same when you reach to touch its ear and be surprised when it shies away. If that happens, back off. Wait until its head comes down, and give it grain with one hand while touching the side of its head with the other, working your way up until you can approach its ear. Touch the outside of the ear a little at a time, and if the horse stands for that, back

off and give a treat (**Figure 4**). Do it a little at a time and don't be in a hurry. Eventually you should be able to get your horse used to running your hand over the back of its ears, squeezing the two sides gently together, and running a finger inside the ear.

After the head, start working down the neck to the shoulder of the horse, touching everything with your hand. When you run your hand over the forequarters, use one hand braced against the horse's shoulder and the other one stroking it (**Figure 5**). If the horse pushes against you, your braced hand will keep you an arm's length away from being run over. If you don't have a hand braced there and you are too close when the horse spooks, you could fall under the horse. If your horse does push into you, move with it and keep a safe distance away. Don't allow it to crowd you. Finish one side on the horse before you go to the other side. Never step between the horse and the wall.

When you get to the rear quarters of the horse, use the training stick for safety. First introduce the stick to your horse; the horse should see it as an extension of your hand. Beginning with the forequarters, work your way to the hindquarters, touching everywhere with the stick, including the belly, up and down the legs, between the hind legs, the back, and the tail (**Figure 6**). Remember: when you touch the horse with the stick, the other hand should be braced against the horse. Treat the horse with grain every time it stays relaxed and doesn't react. You should give a treat only when there is no movement from the horse. Don't keep poking the horse if it is moving and resisting. Stroke it rhythmically with the stick, like petting it, and tell the horse to relax until it doesn't react to it or move. Then give the treat.

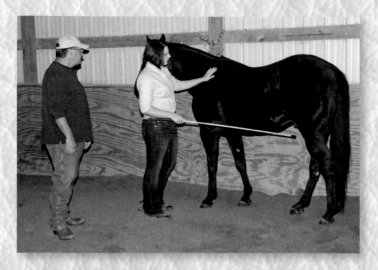

Figure 6. An owner touches the body of her horse with the training stick.

Figure 7. This is how a horse looks when it is afraid. It is pulling back, its head is lifted, and its eye is large and round.

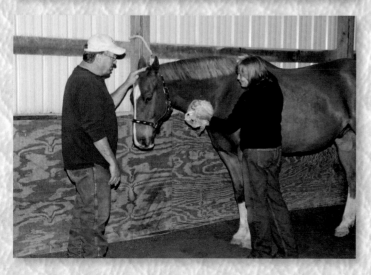

Figure 8. An owner introduces a scrunched-up plastic bag to her horse.

Figure 9. The owner rubs a plastic bag on her horse's ear and poll.

The main thing about despooking is not to get ahead of the horse. Notice if your horse has a relaxed look or not when you introduce each item. If your horse is chewing and has a relaxed look in its eye, it's ready to be touched. If the eye is bigger and rounder and the head is elevated, it's not ready (**Figure 7**).

If you have children helping with treats, you have to tell them when to treat because they can't see what you are doing if your back is to them. Having someone else give treats is helpful because you will have one hand braced on the horse's side and the other one holding the stick and stroking the horse. Give treats whenever the horse is calm and doesn't react to being touched.

As soon as you can touch your horse all over with the stick, introduce the saddle blanket. Be sure to introduce it first to your horse before touching the horse with it. Then touch it all over the horse's body the same as the stick. Though your horse may be familiar with it, it's important to start out with simple objects like this and work your way up to the spookier ones.

After the saddle blanket, introduce a rolled up plastic bag (**Figure 8**). Use either a plastic shopping bag or a white garbage bag. When you introduce it to the horse, bring it to the horse in one hand to let the horse sniff and look at it. Give grain with the other hand at the same time. Keep the garbage bag as small as you can and expect that the crinkly noise it makes may affect the horse. Start at the head

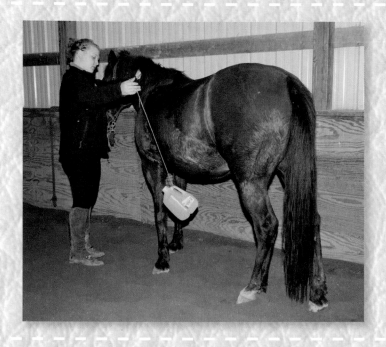

Figure 10. An owner swings an empty milk jug against her horse.

and see if you can touch the bag all around the face and above the head. Only when the horse is relaxed and accepts the bag should you give a treat.

If the horse is sensitive with its ears, do a lot of touching with the bag around its face until it relaxes. Then try to touch an ear. If the horse gets too skittish, retreat. That would indicate you did too much or went too quickly for the horse to handle. When it seems to be more relaxed, try to pet its face and neck with the bag. If the horse tenses up or is even the slightest bit hesitant, back off and wait until it relaxes. Then start over slower and do just a little bit at a time.

If the horse still gets a little touchy the farther up its ear you go, start at the bottom of the ear, then withdraw and give a treat. Start again at the lower part of the ear (**Figure 9**) and go a little farther up, then withdraw and treat. Repeat this process until you can go all over the ear with the bag.

When you get to the horse's shoulder, to stay safe, have one hand braced on its shoulder and try to touch the sides and belly with the other holding the bag. Don't give a treat until the horse is relaxed and has absolutely no reaction. If it tenses up or quivers when you touch it, don't give a treat. You can let the bag get bigger when your horse stays relaxed.

Once the horse is okay with each item, it's time to work on movements and noises. If the horse reacts negatively to a quick movement or loud noise, the wrong thing to do is continue at the same level. Decrease to the movement level the horse will accept, like doing a little jump or just flinching, and give the horse grain when it doesn't react.

To find the horse's threshold for noise, stand a distance away from it, make a loud explosive noise, and repeat it at progressively reduced octaves until you find the level the horse is comfortable with. Repeat the noise the horse will accept while giving it a treat. Eventually you will be able to increase the level of noise and agitating movement. Going to the milk jug later in the lesson should help.

The more time you spend doing these steps, the more you will get your horse desensitized, even if it's just a little at a time. This isn't a one-time cure.

Now we move on to the milk jug. Use a gallon-sized, empty, plastic milk jug. Introduce it to your horse first just like the plastic bag. Bring it up to your horse's face while giving a treat at the same time. Then touch it all over the horse's body beginning at the front end of the horse. Tie it to a length of baling twine and swing it all around the horse's body, touching the horse with the jug (**Figure 10**) just as we did with the blanket and plastic bag, until the horse relaxes. It's lightweight, so it won't hurt.

SUMMARY

If you discover that your horse has an issue with any of these objects, movements, or sounds, you are doing great! Just stop and fix it. For instance, you might find your horse is fine with everything except it doesn't want to be touched on its belly. Taking time to fix this is something to be taken seriously. If you are out on the trail and your horse is touchy with its belly, it could be dangerous. One lady had a horse like that and during a ride some brush got caught in her stirrup, but she didn't know it. When she nudged her horse with her heels, the brush poked the horse in his belly. He spooked, dumped her, and took off for home, miles away.

We are making progress desensitizing our horses so they will be safe to ride anywhere, and we are having fun at the same time!

> **Total time for this session: 25 minutes**

Dennis directed these exercises:

1. Taught the class how to desensitize their horses to being touched all over using their hands, a training stick, a saddle blanket, a plastic garbage bag, and a milk jug.

2. Helped the owners with various problem areas on their horses.

A GIRL AND her Gypsy Horse have fun in a costume class. Willingness is a consistent quality of this gentle breed. Their go-with-the-flow personalities are what they are bred for. Their beauty is an additional perk.
Mammano Photography

Figure 1. An owner introduces the tarp to her horse by bringing it up to its face and treating the horse at the same time.

PART II: TARPS

I have a lot of fun helping owners desensitize horses. The main thing to remember is that with all the different horse owners with different levels of horse experience, the idea is the same: Go slow, don't get ahead of your horse, stay safe, and have fun with the horse. If you can't touch the horse with your hand or the stick, you obviously don't want to go on to bigger and spookier things like tarps just yet.

Everyone attending this lesson has gone through the previous process with their horse, from touching all over the horse's body with their hand, then the stick, to plastic bags, and finally milk jugs. Now we move on to the tarp, which is like a gorilla from the horse's point of view. Tarps are always flapping and noisy, especially if there is a wind, which also makes horses nervous. Trail riding when it's windy can be tough because wind moving the trees looks like there is a mountain lion up there to horses and they freak out. When there is a wind blowing and we are doing training, it's a whole different scenario than when it's calm. So a tarp flapping as if the wind is blowing and making noises at the same time is particularly spooky to horses.

Horses have great vision and can literally see something a hundred yards away. If it's scary, they'll freeze like a dog just before they

Figure 2. I show an owner how to brace against the shoulder of her horse with one hand and bring the unfolded tarp up to it with the other.

Figure 3. I shake a tarp behind the line of horses and keep an eye out for any horse that isn't handling it well.

turn and flee. Something behind them might startle them also, so we're going to desensitize both in front and behind each horse with the tarp and lots of movement and noise.

THE LESSON

I direct the owners to take a tarp and scrunch it up as small as they can get it. Introduce it to the horse the same as we did with the other items in the previous despooking lesson. (**Figure 1**). If the horse pulls back when the tarp is brought up to its face, the owner should draw the horse in as we did in the roundpen by stepping back and telling the horse to relax. Don't show the tarp to the horse if it is at the end of the rope. Step back and see if the horse will

come forward a step or two. If the horse does, treat it with a small handful of grain right away. Then introduce the tarp again and treat the horse at the same time.

Each owner brings the tarp up to its horse and strokes its head with it.

When the horse accepts the scrunched-up tarp there, start opening it up and touching it to the horse's body. You can open it progressively bigger, but take your time. It's easy to unwrap, but start with it small. Desensitize both sides of the horse, but don't go to the other side of the horse until the first side is done.

Be sure when you are using the tarp to have one hand braced against the horse in

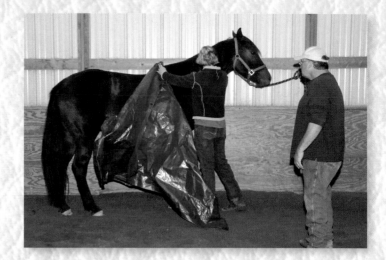

Figure 4. An owner brushes the tarp on her horse to desensitize the horse to it.

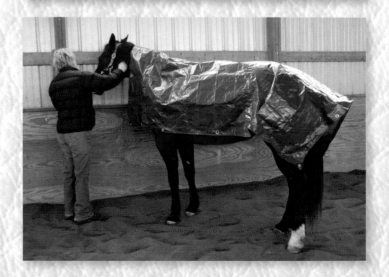

Figure 5. An owner inches a tarp up toward her horse's ears.

Figure 6. A horse stands calmly with a tarp draped on it like a blanket.

case it spooks and slams into you (**Figure 2**). Don't hold a treat in your hand. Either have a treat master with you or return to the grain bucket yourself for each treat. If you have the tarp in one hand and a treat in the other when the horse bumps into you, you've got no recourse but to fall on the ground and get run over. However, if you brace with your free hand and the horse smacks into you, you are balanced and can push away, so don't hold a treat in your free hand. Otherwise, the horse could end up on top of you. It's kind of like dancing: If the horse pushes into you, go with the flow and do what makes sense.

For this part of the lesson, I am going slowly up and down the line of horses, gently shaking an opened tarp, moving it around, and waving it behind each horse about ten or fifteen feet away (**Figure 3**). The owners step aside while I do this in case their horses get too excited and slam into them, but they step back again when their horses accept it. They continue with their own tarps, simultaneously allowing their tarps to get bigger and flapping the tarps at their horses as they accept it more. So the horses are getting a double dose of tarp movement, both from their owners and from me.

The nice thing about this part of the clinic is that each person is desensitizing their neighbor's horses while doing his or her own horse. It's fun to see people helping each other by training their own horse.

If I see a horse reacting badly as I go up and down the line shaking the opened tarp, I'm not going to do anything but wait until it calms down. I may step away at a greater distance before shaking the tarp again or just drag it along the ground, but we'll work on it until the horse calms down.

It's kind of like dancing: If the horses pushes into you, go with the flow and do what makes sense.

I warn the owners to get their hands braced against their horses because I'm coming closer behind them shaking the tarp. If a horse moves against an owner to get away from the tarp, the owner needs to be able to move with the horse. I remind them to give a treat to a horse that doesn't react as I flap the tarp. If the horse turns away, try to draw it back in.

I keep flapping my tarp, and some horses have a reaction and some don't. I try to flap it when a horse has its side next to the wall and the owner is in the safe area on my side. I warn the owners not to go on the other side of their horses when I go by flapping the tarp. If the horse even flinches, I take it real easy and slow. The owner should be ready with the treat as I pass by shaking the tarp.

The owners are concentrating on getting their own tarps bigger as they work on their horses. Some are brushing their horses with the tarp (**Figure 4**). Others are smacking it against their horses' bodies back and forth from the ground to their sides. Some are swishing it from the ground to their bellies. The tarps are lightweight so they don't hurt, but I warn the owners not to snap it like a rolled-up towel.

I walk down the line again while smacking a milk jug against the tarp I'm holding. I quit when I'm by a horse that looks nervous.

Don't be discouraged if your horse takes longer to desensitize than others.

I wait until I'm on the owner's side of the horse to continue so it won't bump into the owner. With a nervous horse like this, we do one tarp at a time: I do mine first and then the owner does his. His horse will need more work, but the owner is doing great with desensitizing it.

An easily spooked horse is a dangerous horse. If I were on a trail ride with a horse in front of me that gets fearful like this one does and I were the same distance away from it as I am now, that isn't much of a gap between us for it to spook and get out of control.

This is a team effort when I shake the tarp from a distance and the owner is doing the same thing up close to the horse. If a horse doesn't react, that's the time to give a treat. If it even raises its head, it is reacting and shouldn't be treated. When it accepts all the commotion, give it a treat.

If the horse is ready, meaning it has fully accepted the opened tarp on every area of both sides of its body, then throw the tarp up on its back the same way you throw on a saddle. Don't do it too soon if your horse isn't ready. If the horse stays calm, then treat the horse. Position the tarp like it's a horse blanket and pat all around on it so it makes noise. Treat the horse at the same time if it stays relaxed. Really flap it a lot around your horse.

Next we move to the other side of the horse. Pull the tarp off and carry it around to the other side, but not too close to the horse. If you are walking too close around your horse carrying the tarp and the horse spooks and kicks at the tarp, you will get nailed. So walk a good distance around your horse to be safe. I've had more milk jugs kicked out of my hands by walking too close than anyone else.

The next step is to pull the tarp up to the horse's head; don't move on to this step unless the horse is really ready. Bringing the tarp up behind its ears is as close as you want to get for now and do it a little at a time (**Figure 6**). If the horse shies, withdraw whatever you are doing and do it again inch by inch, just like saddling a horse the first time.

If your horse is doing well and has accepted everything you did with the tarp, then give it a treat and take a break. It's better to take a break than overdo it. The owners on either side of you might still be despooking their horses, which will continue to affect your horse when you leave, so your horse will still be learning. Though your horse might be noticing its neighbors, if you have done the despooking the right way, it should be fine. A good sign that a horse is relaxed with all the commotion going on is if it is chewing.

I walk on to another horse and as I shake my tarp, the horse puts its ears back and seems a little disturbed. I back up and do it from a greater distance. The owner flicks her tarp at the horse's legs, under its belly, and the space all around the horse until it is relaxed. As the horse relaxes, the owner treats it.

More horses are relaxing and their owners are taking a break. The horse that was the most skittish is now standing quietly with a tarp draped over it like a horse blanket (**Figure 7**). The horse is calm as the owner pats all over the tarp making crinkly noises. As I shake my tarp about ten feet away, the horse settles down and doesn't react anymore, so the owner gives a treat while I am still shaking it. This was the last of all the horses to be desensitized to everything we did with the tarp.

SUMMARY

Don't be discouraged if your horse takes longer to desensitize than others. Every horse is different, and the next time it may be the other way around. The point is to take your time and have fun. Don't get frustrated or get ahead of your horse.

I'm a boisterous talker. Sometimes when I go into other stables where people are working their horses, they ask me to be quiet. They say I am scaring their horses, but the truth is that it's good for horses to be exposed to different noises and sights. They have to get used to their environment one way or another. The despooking we are doing is helping our horses to relax no matter what noise is going on, which makes them safer animals to be around.

The horses need to take a break and think about the lesson now. I'll go to the next step after the break and bring everyone back to their horses.

| Total time for this session: 14 minutes |

Dennis directed this exercise:

1. Helped owners desensitize their horses to the tarp from behind as well as all over their horses' bodies.

AN ORIGINAL American Indian horse is encouraged by her rider to maintain slow and steady forward motion through a stream, overcoming the mare's natural urge to stop, play, or even roll because of her love for water. *Chris Hurd*

PART III: LOUD VEHICLES AND CROSSING TARPS

The next phrase of my despooking clinic steps it up a notch, moving outside for the obstacle course. I'm working with the same horses from the previous despooking lessons (see pages 139–153) and coming back to them on a super positive note. All the horses are cool, calm, and relaxed; they're not worn out.

LESSON, PART 1

I ride an ATV four-wheeler up to the line of owners and their horses (**Figure 1**), then stop and let it idle. Horses will spook with ATVs and other running vehicles, so we want to make sure our horses are traffic safe.

The owners bring their horses up to the front of the idling ATV and introduce it to their horse the same way we did with the tarps, jugs, and other spooky items. I have the grain bucket with me and will treat each horse when it comes close enough.

The owners have their horses lined up in front of me, and I direct them to come one at a time and stay well apart from each other in case a horse kicks. They approach slowly, come right toward me, and allow their horse to smell the vehicle while it is idling.

The first horse stops in front of the ATV, looks at it, and snorts, which means he's a little reluctant, but the owner is able to walk him around to the side close to me (**Figure 2**). I let him smell my hand and he relaxes, so I give him a treat. We give him all the time he wants to check out the vehicle. Then the owner leads him past the ATV.

The next horse walks straight up to the ATV with no issues, so I give it a treat from my seat. The next horse approaches, smells the vehicle (**Figure 3**), and takes its treat from me. I give the owners some grain to treat their horses as they pass the vehicle.

Each horse needs to be introduced to the vehicle because it's new to them. All of them want to smell it. When the calmest horses approach, I make some explosive noises and give them a treat at the same time to desensitize them more.

The next owner gives a treat to her horse as it smells the vehicle, but she needs to let her horse relax more to bring it up close enough to get a treat from me. The horse takes another step forward and gets a treat. A young girl is with the owner and I direct the girl to go the other way around the vehicle instead of close to the horse for safety reasons.

Each horse is introduced to the ATV and lines up with about twenty feet between them. I have them come back again one at a time to see the ATV from the backside (**Figure 4**) instead of the front and give each a treat as they come close. The first horse brought forward is the one that is the most nervous, and I coax the horse closer with a handful of grain and it passes by quietly.

I caution everyone not get on the far side of their horses as they approach the ATV. Never get on the opposite side of your horse when it is focusing on something spooky. If the horse freaks out, you don't want to get knocked over as it turns and runs away from what spooked it.

Each owner gets a treat from me as they get close to the rear of ATV. With the calmer horses, I again make a noise similar to a gun to desensitize them to loud noises.

As everyone passes with their horses, they line up again and face me. I remind them to have their bucket of grain close by for treating because I am going to drive by on the ATV.

STEP-BY-STEP

Figure 1. I drive the ATV up to the line of owners and their horses that are well spaced apart.

Figure 2. The owner manages to get her horse to the side of the ATV, and I encourage it in with a reward.

Figure 3. An owner brings her horse up to the ATV to check it out, and the horse sniffs it.

Figure 4. A horse calmly accepts the ATV at the back of it while another waits its turn.

Figure 5. I pause and rev the engine with the ATV close to a horse while the owner gives it a treat.

Figure 6. I straighten a tarp while helping an owner introduce her horse to the tarp. A grain bucket is handy to have for treating the horse.

Figure 7. I help introduce a nervous horse to a loosely piled tarp. The tarp is similar to the one that was rubbed on the horse's body previously to desensitize him.

Figure 8. The horse that wouldn't cross an opened tarp before can now be walked over it. The owner is happy with the horse's response.

Figure 9. I flap the tarp as I drive the ATV slowly around the horses.

I'm going to get closer with repeated drive-bys, and they are to treat their horse when it remains calm.

I drive slowly about ten feet away from the horses. As I get close to each one, I direct the owners to give their horses a treat. The ATV makes a different noise as I slow down by each one, but the horses accept the noise. I drive back and come closer to each one and rev up the engine as I pause near them (**Figure 5**). They are all calm and accept the noisy ATV.

LESSON, PART 2

Now we're going to do something different with a little obstacle course on the ground. There are five open tarps spread out in a large circle on the ground. I stand beside the first tarp with a bucket of grain (**Figure 6**). I am the treat master this time, but there are grain buckets spaced out by the other tarps also.

I direct the first owner to introduce his horse to the tarp: Let the mare smell it and then walk her over it. The owner stops the mare when she is standing completely on the tarp and gives her a treat. Once she walks over it, she gets another treat. I want each owner to do the same and lead their horse over every tarp and go slowly.

The next owner approaches and introduces her horse to the tarp on the ground. The horse steps on it and gets a treat, so the owner moves on to the next tarp. I direct everyone to stay at least three horse lengths away from each other.

The nervous horse is next so I help with him. We show him the tarp and let him sniff it, but he doesn't want to step on it. I scrunch it up to make it look more like the one rubbed on him in the previous lesson (**Figure 7**). For some horses that absolutely refuse to step on an opened tarp, it can be folded smaller or scrunched up a little so it isn't as intimidating. Rewarding horses like this will be important. The tarp can be gradually opened as they gain confidence until they will walk on it when it is flat.

I have to coax the nervous horse with a little grain; he puts a front foot on the tarp, then steps on it with both front feet. I stop him there and let him relax and have a treat. Front feet are easy; it's the back feet I worry about. Unlike many other animals, horses can't see their hind feet, so owners have to stay way off to the side in case the horse spooks and jumps sideways. But the horse steps on the tarp with his hind feet, and I am able to walk him over it. He gets a treat. The same horse approaches the next tarp and only hesitates a little before walking over it (**Figure 8**).

All the owners are walking their horses in a large circle over the tarps. There are lots of buckets of grain spaced at intervals, and they stop and give their horses treats as they walk their horses over the tarps.

Next I warn them to be prepared as I start up the ATV and ride it down the middle of the circle, stopping at the first tarp. I drive through the circle and in between two of the horses, then stop the ATV and start it again several times.

Owners continue walking their horses over the tarps as I drive around and through the circle. I pick up one tarp and wave it as I drive around the circle. I flap it a little, going slowly around the more nervous horse. I stop now and then and flap the tarp off to the side of the ATV (**Figure 9**).

As two horses pause between two tarps, I warn the owners that I am coming between them. There is a ten-foot gap between them and I drive through, but the horses stay calm. I turn around and warn everyone that I am coming behind them, and after I drive by, I swing around and stop in front of each horse, revving my motor. I drive both fast and slow between and around them.

I have the owners line up their horses as I drive fast circles around each horse and they all handle it well. Everyone did a great job, and we are going to take a break.

SUMMARY

Because we continue doing the despooking the same way we started, the horses are calm, expecting to get treated with each new object we present to them. If we had started with an ATV vehicle revving up its motor instead of touching the horses all over with our hands (see page 142), the despooking lesson wouldn't work.

People who attend my clinics say it is so cool to be doing all this despooking in a few hours, but keep in mind that one time is not a fix. We have to keep desensitizing our horses. Each and every day, try something different, unique, or challenging to you and your horse, but don't get ahead of the horse. We want to make sure our horses are desensitized with all these objects before we mount up and ride.

Our horses are going to be cool, calm, and collected the majority of the time if we don't rush through it. So just relax and enjoy it!

> Total time for this session: 15 minutes

Dennis directed these exercises:
1. Desensitized the horses to an ATV driving close by and revving its motor.
2. Desensitized the horses to walking on tarps.
3. Desensitized the horses to flapping tarps on an ATV.

A STATELY 17.2-hand Shire stallion warms up before leading a draft horse parade in northern Minnesota. Riding on the substantial back of a huge draft horse like this with its powerful, rolling stride is truly an unforgettable experience.
Richard B. Hicks, Russ Brand Peridot Shires

PART IV: STREAMERS AND FANS

By the fourth part of my despooking clinic, the horses have been walked over tarps and up to an ATV and are getting more confident as their owners introduce each new item. We've started a good mindset with them to accept spookier things. For this next segment of the clinic, the horses are now saddled and carrying grain for treats on a horn bag on the saddles. They are still haltered but not bridled yet.

A large barn fan is set up with bright plastic streamers tied to it that blow and flap when the fan is turned on. (Alternatively, plastic bags can be used.) To a horse, the blowing plastic is really spooky.

THE LESSON

I ask each owner to walk up one at a time with his or her horse and have the horse face the fan. Before approaching the fan, they walk over a tarp that is about twenty feet away. After the horse's introduction to the fan, I want each owner to stop the horse alongside the fan as they pass by it and then treat their horses there (**Figure 1**). The more each horse sees the fan and streamers, the more desensitized it will get. I position myself nearby to help.

If a horse stops a distance from the fan, that's all the farther the owner should go. The horse should be given a treat there and given a chance to relax before going forward. I caution owners to stand between the fan and the horse because if the horse spooks, it will jump away from the fan; if the owner is in the way, they could get run over. Also, if the owners are standing too close to each other, they might get kicked. A telltale sign

that a horse is starting to spook is when its eyes get big.

The first horse and owner approach, and the horse stops about ten or fifteen feet from the fan, so the owner gives it a treat there (**Figure 2**). When giving the horse a treat, have the horse reach down for it so it relaxes, then try to move the horse a little closer. The owner gets the horse to move a couple of steps closer and waits until the horse puts its head down to treat it. When she is able to get the horse closer, she treats it again.

The second horse and owner approach. When the horse stops to investigate, the owner stops there and gives her horse some grain. The horse is totally engrossed with looking at the fan and the flying streamers. This is a spooky thing to a horse. As the owner leads her horse past the fan, it hesitates and observes the fan intently. She stops there to let her horse look at it and gives a treat, then moves on. As she does, she treats her horse again.

The next owner comes with her treat ready in hand. As soon as her horse slows, she stops. The horse snorts, which is a sign to stop. When horses snort, they are saying, "What is it?" Then they'll suddenly brace just before they explode. So as soon as the horse snorts, that's a good place to stop and give it a treat. The horse steps up on its own to get closer to the fan, which is great. With its ears pointed at the fan, it is curious and that is another good time to treat it. I tell the owner to be sure to give her horse a little more leadrope so she isn't too close and can step away quickly if she has to.

Before each owner brings their horse up to the fan, they walk it across the tarp. After introducing their horse to the fan and when walking past it, I want each owner to stop when their horse is parallel with the fan and give it

Figure 1. An owner walks her horse toward the fan with a treat in her hand.

a treat there because we are desensitizing the side of the horse to the fan as well (**Figure 3**).

The next horse is more cautious and walks slowly up to the fan. I caution the owner to stop a good twenty feet away and reel in her lead line a bit more. The horse is curious and as long as it is not being forced, it walks up willingly to within five feet of the fan, then raises its head. That's another good indication to stop right there because the horse is getting alarmed. If the horse wants to back up a little, that's fine, but the horse's comfort level is right where he stopped, so the owner treats him there. If we tried to force a horse like this to get close to the fan all day long and it didn't want to, it would just get worse.

The owner tries to bring her horse in a little closer, but I warn not to pull on the leadrope. I kiss to the horse and he slowly takes another step. I tell the owner to make her horse put his head down for the treat because it will relax him. I demonstrate by raising my hand in front of the horse's face and he raises his head to follow it. When I bring my hand down in front of his head, he follows my hand and his eyes soften as he brings his head down. When a horse is relaxed like that, it gives us more control and he is safer to be around.

As the owner leads her horse by, it stops to look at the fan again, which is great. Allow the horse to look until it is satisfied and treat it there.

Figure 2. When a horse stops to look at the fan, the owner is ready with a treat.

Figure 3. The owner treats her horse again when it is parallel with the fan.

The next horse comes, and the owner gives it a treat when it stops. I demonstrate to her about noticing the softness in her horse's eyes and how they change if the horse gets nervous. When the horse drops its head to get a treat, the horse relaxes and its eyes noticeably soften. Until a horse does that, don't force it any closer to the fan. When its head is up, the horse is too alarmed to go forward. Give the horse a treat when its head is down.

The owner is now able to move her horse closer and gives it a treat. As she walks her horse past, she stops it alongside the fan to desensitize the side of her horse. I direct her to get in front of her horse so it can see the entire fan and the streamers blowing from it.

The horse is calm, so the owner gives it a treat and walks on. Like all of my lessons, this one involves repetition. You want to get the horse used to whatever you are teaching, and that doesn't happen from one try.

I usually teach owners to have a long lead-rope on their horse, but in this case, have a shorter grip on it (**Figure 4**) because if a horse spooks, the owner can keep the horse off of himself or herself better if there is less slack in the rope.

The first horse comes up to the fan again. With its head slightly elevated and its eyes round, the horse is showing signs of nervousness. I tell the owner to stop right where he is and treat his horse. Then he is able to walk

Figure 4. Hold the leadrope close to the horse's head to keep control of the horse in the event it tries to move too close.

Figure 5. Offer the treat low so the horse will bring its head down for it. This will relax the horse.

Figure 6. I rev up the ATV next to a horse by the fan.

it a step or two closer and gives it a treat. I show the owner how to bring his horse's head down to really smell or look at the fan. The owner walks by and stops with the horse at an angle to the fan. I gradually turn the fan on the horse so it can feel it; the horse stands well and the owner treats it.

The next horse is already stopped and observing the fan a distance away with its ears forward and its head slightly elevated.

I tell the owner to let her horse relax first before trying to come forward more, but the horse stalls in that spot. Here's a little trick: Get some leverage on the horse by stepping to the side and directing the horse to turn its head and step aside with you. I try to draw the horse to the left side of the fan, but it won't move. This is why it's important to work on both sides of a horse. Coming to the fan on one side is not a problem for this horse, but the other side is.

It's great if owners can approach the fan with their horses straight on, but if the horse is a little too nervous, try to give the treat low so it will put its head down. I show an owner how to do this (**Figure 5**) and the horse's eyes soften as it brings its head down for the treat. We are able to get the horse closer and the owner leads it off to the side and stops the horse there to treat it again. Every time she brings her horse up to the fan, she should do the same thing: Give the horse a treat low so the horse brings its head down and relaxes.

The next horse comes and the owner introduces it to the fan for the second time. The horse is getting really close to the fan now and sniffs the streamers. I caution the owner to be careful where she stands. A horse can go six ways when it spooks: up, down, sideways each way, forward, or backward. The owner gives the horse a treat, and it seems quite

One of the reasons our horses are more sensitive on one side than the other is because we don't spend enough time on their right side.

relaxed. As she brings the horse to the side of the fan, it sniffs it.

The next horse approaches with its head up, so I tell the owner not to treat it there. I back the horse up a step, and its eyes soften. That is the spot to treat it. One step closer and the horse was out of its comfort level. When it raised its head and pointed its ears at the fan, it was too nervous, but as soon as it backed up a step, the head came down. With the horse more relaxed, the owner tries to bring it forward and it easily takes a step closer, so the owner treats it.

In order to get both sides of their horses desensitized, the owners are leading their horses off to the other side of the fan now. When they stop close to the side of the fan, I direct them to get in front of their horse and out of the way so their horses can see and feel the fan before they get a treat and move on.

The next two horses are calm and get quite close to the fan. One gets close enough to touch the streamers. When the other horse passes the fan, it turns to sniff the side of it and the owner treats it. All the horses are getting used to the fan now.

I have the owners continue to lead their horses up to the fan while I turn on the ATV and ride it in and around them (**Figure 6**).

As I get close to each horse, the owner gives it a treat. The horses are getting closer to the fan and aren't bothered by the ATV running around them because of the previous desensitizing we did.

When each horse is brought up to the fan, I stop close to its side and rev up the motor. The owner gives a treat at the same time. Then I ride behind them as they face the fan. They all do well. I also ride up closer to them when they are facing the fan. They are all getting really close and desensitized to the fan by now.

SUMMARY

It's time to take a break and everyone ties up their horses to let them relax and think about what just transpired.

One of the reasons our horses are more sensitive on one side than the other is because we don't spend enough time on their right side. We mount, dismount, start grooming, and do everything on the left side instead of both sides. Keep this in mind when you are despooking, and always work the lesson from both sides.

Also remember that you can learn a lot about your horse by watching its eyes and ears. When a horse has its eyes and ears going forward, it is pointing at something. Horses will do that on a trail ride if they see a deer at two hundred yards out; the rider may never see it at all but realize the horse is pointing at something. We can learn so much just by watching our horses, and this will help when trying to despook.

Total time for this session: 16 minutes

Dennis directed these exercises:
1. Desensitized the horses to a barn fan with plastic streamers attached and blowing out.
2. Desensitized the horses to the ATV driving around them and revving up next to them as they are brought close to the fan.

A RIDER on his Spanish Barb demonstrates mounted shooting at an exhibition. Barbs are especially adept at this event because of their good minds. They run the course with confident composure—no head tossing or frantic looks, just quiet perseverance.

Marjorie Dixon and Maureen Kirk-Detberner, Fast Winn Photo

STEP-BY-STEP

Figure 1. Owners rub plastic bags on their horses' foreheads to distract them. (At the beginning of the lesson, the shooter has to be a greater distance away.)

PART V: GUNS

With each step in the despooking clinic, the lesson gets more difficult, which means it takes more time. We've already introduced our horses to many scary things and now we're escalating it further. We are actually going to start shooting a gun around them.

I don't advocate violence, but the reason I came up with this lesson is because I used to have an old lawnmower that would backfire every time I turned it off. It scared every horse in the neighborhood. I thought shooting a gun would simulate the backfiring of a motor or something similar, like firecrackers going off. If you ever ride in a parade, you will know how important it is for your horse to be familiar with loud bangs and noises. Gunshots

or vehicles backfiring can be heard even on trail rides.

THE LESSON, PART 1

All the owners in the clinic are standing next to their horses and lined up. They are well spaced and facing the opposite side of the arena. I have my friend, Ben, standing some twenty-five to thirty yards away from the horses with a handgun. I am standing off to the side so I can watch what the horses' reactions will be.

This is what we are going to do: First, the owners are going to rub a plastic bag on their horses' foreheads, which the horses won't react to because they have already been desensitized to plastic bags. As they rub the bag on their horses' foreheads, I'm going to shout, "Ready!" At this point, they will start

Figure 2. A horse lifts its head at the sound of the gun, so it isn't allowed to move forward.

Figure 3. Two horses step forward toward the shooter.

rubbing the bags extra fast on their horses as a diversion. Then I'm going to count to three, and Ben will shoot a blank up in the air.

The owners will handle their horses depending on how their individual horses react. If the horse doesn't jerk or react to the shot, the owner can lead it five steps forward toward the side of the arena where Ben is standing and give it a treat. If the horse reacts a little, they can take only a step or two forward. If the horse reacts a lot, they stay where they are.

Everyone will be moving their horses closer to the gun as they get desensitized to it going off, but they won't get a treat until after they have stepped forward, closer to the gun. This technique allows horses to get desensitized to gunshots gradually at their own pace.

I came up with this exercise because it took forever to desensitize horses to the sound of gunshots without distracting them with something else. Rubbing a horse's forehead with a plastic bag is a relaxing distraction and the horse gets a treat if it stays calm when it hears the gun. It's a quick and easy way to get horses used to gunshots.

I want to emphasize this is not a race to see who can get to the other side the fastest. A lot of people, especially kids, want their horses to be first to the finish line and "win," but we need to stop and think about this: The longer it takes to reach Ben, the more desensitized our horses are going to get. Those that reach Ben the quickest will hear the gun the loudest. They will have to be really calm and accept the sound of gunshots. Every horse is

STEP-BY-STEP

Figure 4. I help the owner to get her horse's head down with a light touch and a handful of grain.

Figure 5. An owner stands at the side of her horse that is closest to the shooter and covers her ears when the gun is shot. It's loud for the horse too, but we still don't want it to react.

Figure 6. I work with the horse to get its head down.

Figure 7. A horse stands quietly and comfortably by the shooter.

Figure 8. The horse are able to stand close to Ben now as he shoots upward.

The most important thing to remember in the gun despooking clinic is to take your time and follow your horse's own pace. If it adjusts quickly to the bang of a gun close up and all around it, great! If it takes a lot more time, that's great too! It's just more of an opportunity for you to ensure your horse is desensitized to loud noises.

different, and we don't want to go beyond what our horse is capable of handling.

I caution the owners to make sure their leadropes are not on the ground or their horse might step on them and they'll get bumped when the gun goes off, which will scare them more.

I shout, "Ready!" The owners start rubbing plastic bags on their horses' foreheads (**Figure 1**). I coach them to rub really fast and then I shout, "One . . . two . . . three!" and Ben shoots the gun up into the air. All the horses jerk their heads up. Because they all react, they don't get to move forward or get a treat. Only when they have no reaction can they be lead five steps forward toward Ben and get treated. They can have a little reaction, but if they have a strong reaction, there will be no treat.

The owners are still lined up facing Ben and holding their horses by their leadropes. They rub their horses' heads with the plastic bags again, which should relax them and bring their heads down.

I shout, "Ready!" Then I tell the owners to start rubbing the bags on their horses really fast. I count "One . . . two . . . three!" and at three the gun goes off. Some horses lift their heads (**Figure 2**), and others don't. I am

watching and tell each owner if their horse deserves to move forward and get a treat or not. Only one horse can take five steps forward and get a treat.

I remind the owners that this isn't a race. We're going to take our time, especially in the beginning, because the horses need time to adjust to the gunshot. The owners should hold their horses close and try to get their heads down. When I say, "Ready," they are to rub the bag as fast as they can on the horses. It will keep their attention directed away from the gunshot until they adjust to staying calm when they hear it.

I shout, "Ready," the owners rub the bags on their horses faster, and I count, "one . . . two . . . three!" At the count of three, the gun goes off.

Again, the horses have different reactions. Some don't react and can step forward. They move forward according to how much they reacted. One horse can take five steps forward, and another one can take two steps because it reacted to the shot but not radically (**Figure 3**). Both are treated after they move forward.

We are going to move quickly now that everyone understands and the horses are starting to understand too. I shout, "Ready!"

and the owners rub the plastic bags on their horses' heads. I tell them to do it quickly and then count to three and the gun goes off. They all get to move forward five paces this time.

We repeat the sequence: "Ready! (Rub faster, faster, faster!) One . . . two . . . three!" The gun goes off, and some horses lift their heads. Some can take five steps and others can only take a few steps, but they all can move forward and get a treat.

It's important to try to get the horses' heads down when rubbing to relax them. I demonstrate by rubbing a plastic bag on the forehead of a horse that has been reacting to the gun. I tell the horse to relax as I rub its head and apply a little pressure on the lead-rope, pulling it down. The horse drops its head and gets a softer look in its eyes. While I continue to rub, I shout, "BOOM," as loud as I can and there is little reaction from the horse.

The group does it again. When the gun goes off, three horses stay relatively calm and can move forward and get treats. Another one can take only two steps forward because he jerked his head up when the gun went off. Other horses reacted too much and can't go forward.

We have to do this faster because horses get antsy with this lesson if it takes too long. I say, "Ready," and the owners start rubbing their bags on their horses. I coach the owner whose horse is reacting the most to start working on getting her horse's head down. I count "One, two, three!" Bang!

When the gun goes off, two horses stay calm, but the horse that is last can't step forward at all because he jerked his head up and spooked a little, which is too much of a reaction. I have the owner turn her horse around in a circle to get the horse to relax and get his head down. As she walks her horse, his head lowers, and I tell her to stop right there. I show her how to get his head down even more by giving him the treat low. He lowers his head to get it (**Figure 4**). It's not good for a horse like this to be too close to a gun. It will sound louder as the horse gets closer to it, so it's good for this horse to be where he is for now.

We repeat the process: ready, rub, count-down, bang. The gun goes off and the same two horses stay calm and can move forward and get treated. The others are left behind.

I realize that the owner whose horse is last appears to be flinching herself as a reaction to the gun, so I tell her to try not to do that. We send our emotions to our horses, and her reaction is having an effect on her horse, causing him to jerk his head. I remind everyone to stay calm and also to spread out from each other just in case a horse kicks.

Now I direct the owners to turn their horses around and to begin again where they all started. We're going to bring Ben in closer to the middle of the arena. As the horses progress moving toward him, they can just pass him by.

The horses are lined up facing the opposite side of the arena. "Ready!" (Rub, rub, rub!) "Set!" (Faster, faster, faster!) "One, two, three!" The gun goes off, and this time they all react. That's not a surprise because they are back where they first reacted.

We are speeding it up now. "Ready! One, two, three!" Everyone did okay and can move forward, but only one step. "Ready! One, two, three!" Two horses are fine and can move forward. We do it again and one horse moves forward. The next time, the last horse has the most reaction, and the owner turns it around again and stops when its head is lower, but it continues to jump when the gun

STEP-BY-STEP

Figure 9. Ben shoots the gun upward about five feet from the side of the horse's hip.

Figure 10. The owner places a hand on the horse's shoulder to brace it there and practices shooting over the horse's back.

Figure 11. The owner shoots a gun off her horse's back.

goes off. The other two horses in the lead are even with Ben now.

I remind everyone to be cognizant of where they are standing by their horse, especially as they get close to Ben. Where is the horse going to jump if it gets scared? It will jump away from the gun, so don't stand on the opposite side of the horse. Stand in front of it toward the shooter (**Figure 5**).

All the owners are working on getting their horse's heads down, and the next time the gun goes off, most of them can move forward. The two lead horses have passed the shooter because they had no reaction to the gun even though they are close to it. I congratulate their owners because their horses have graduated! One of these horses was last in all the previous lessons but is the best at this one. Remember, it's not a competition. Each horse learns at its own pace.

The other horses are almost even with Ben, so most of them are moving along quickly now. I remind the owners not to feel bad if their horses are still reacting. They shouldn't think they are losing. There are always some horses in every class that take longer to adjust to gunshots. They are still way ahead of horses that have never been exposed to guns.

The next thing that happens is a surprise: I help the owner whose horse is last and count, "One, two, three." The horse flinches, but the gun didn't go off. Ben didn't realize he had to reload, so he wasn't able to shoot, but the amazing thing was the horse was expecting it and flinched when I counted to three. So we do it again, except I tell Ben not to shoot and then I count to four. The horse doesn't flinch. The owner and I keep working on getting its head down (**Figure 6**).

At this stage, it's important to move along quickly so the horses don't get discouraged. On the count of three, the gun goes off and the remaining horses have only slight reactions. Though most of them are really close to Ben, they are all able to move a step forward and get treats, including the nervous one that I am helping.

I remind everyone to rub their bags really fast and when the gun goes off, two can move forward and get treated. I count again right away; this time when the gun fires, they can all move forward.

The horse that was the most nervous in the previous lessons is now the calmest one with the gunshots, and the horse that was one of the calmest horses in the other exercises is reacting the most to the gunshots. We can never be sure what a horse's reaction to new sounds and sights will be, even a normally calm horse. That's why despooking is so important.

The horses that have already passed Ben have turned around and are standing close to him, getting super desensitized (**Figure 7**). We are desensitizing everyone simultaneously, but the ones that have already passed Ben are basically done. They are cool, calm, and collected.

We do it again, and the horses that are now in the lead are having reactions because they are closer to the gunshots. The one that has been last and that I have been working with has to be turned around to get his head down because he is still reacting. I counsel the owner not to get discouraged. Nobody wants to finish last, and her horse finished first in all the other exercises. The horse is getting antsy the longer we do this, and he needs to get out and move around. The owner didn't do anything wrong, but that is what happens when we take too long doing this exercise. Her horse is just anxious because it is standing too long and knows there will be a gunshot. I suggest to the owner that she take her horse away and exercise him before bringing him back into the class, which she does.

For this type of lesson, we have to continually move forward. In other lessons it's useful to let the horses stop and think about it. With this one, they get antsy if they stand around and think about it too much, then they want to move around. So it's important to do the lesson quickly and keep the horses moving forward.

We do the lesson again with the remaining horses a couple more times and most of them have no reaction and can move five steps forward. I remind the remaining owners to stay on the side of their horses where the shooter is in case a horse spooks. All the horses are progressing with each successive shot and are able to go past Ben (**Figure 8**).

So all the horses have had gunshots go off in front of them, on the sides of them, and behind them as they pass by the shooter. As the last two horses pass Ben, I congratulate them. They did it perfectly and graduated.

THE LESSON, PART 2

Next we're going to get closer and see if we can actually get to where we can shoot from off the horse's back. The first horse that finished is brought up close to Ben, facing him about eight feet away. The owner rubs the bag on its head. Ben shoots the gun up and the horse gets treated because it stayed calm, but we keep it there and don't move it anymore.

Ben walks next to the shoulder of the same horse and stands about ten feet away, then shoots upward on the count of three. The horse doesn't flinch because it has already adjusted to

gunshots going off next to it and it gets treated without moving forward. Then Ben moves to about five feet from the side of the horse's hip and shoots (**Figure 9**). Again, the horse gets treated without moving forward.

Ben does the same sequence on the other side of the horse and it remains calm. Always remember to work both sides of the horse.

The horse is then saddled and bridled, and Ben stands right next to the horse, bracing with one hand against its shoulder. I hold the horse, rubbing its face with the bag really fast. Ben shoots once holding the gun high and pointing it up; the horse is fine and gets treated. Then Ben points the gun over the back of the horse with one hand braced against the shoulder to keep the horse off him if it spooks. He shoots, and the horse stays calm and gets treated. Cool!

Now it's time to try the same lesson mounted. To start, I mount the horse, the owner gets the bag and treats ready, and Ben hands me the gun. I get on the horse instead of the owner for safety reasons—just in case there is a reaction, I don't want anyone else to get hurt. The owner rubs the bag on the face of the horse. I count to three and shoot the gun holding it up at arm's length, and the horse has no reaction and gets treated. We do it again: The owner rubs the horse's face, I count to three, and shoot. The horse remains calm and gets treated.

The other owners can continue practicing this at home. For their safety, they should start twenty-five yards away from the shooter, then fifteen yards, then ten yards—or whatever distance they feel safe with. They will need two partners: one for shooting and one to watch and be on hand for safety reasons.

Only when the horse has no reaction to gunshots that are up close and all around should they allow a gun to go off as close as what Ben just did (**Figure 10**). This is for safety reasons. If there is no reaction from the horse when the gun is shot off on both its sides, in front and behind, and shot over its back, then the owner can try it while mounted (**Figure 11**).

SUMMARY

This lesson might take ten minutes or it might take days. Some horses just need more time to get used to the bang, so take your time and stay safe.

I don't advocate guns or violence, but there is a huge society out there that loves to shoot blank guns off their horses' backs in competition. This is a fast and safe way to teach a horse how to do it.

This lesson also despooks horses for any loud noise, not just gunshots. Fireworks, old lawnmowers, backfiring trucks—anything and everything that makes a loud, sudden bang could spook your horse. That's why it is valuable to teach your horse to stay calm.

The most important thing to remember in the gun despooking clinic is to take your time and follow your horse's own pace. If it adjusts quickly to the bang of a gun close up and all around it, great! If it takes a lot more time, that's great too! It's just more of an opportunity for you to ensure your horse is desensitized to loud noises.

Total time for this session: 20 minutes

Dennis directed these exercises:

1. Desensitized the horses to gunshots from various distances.
2. Desensitized a horse enough that its owner could shoot a gun from its back.

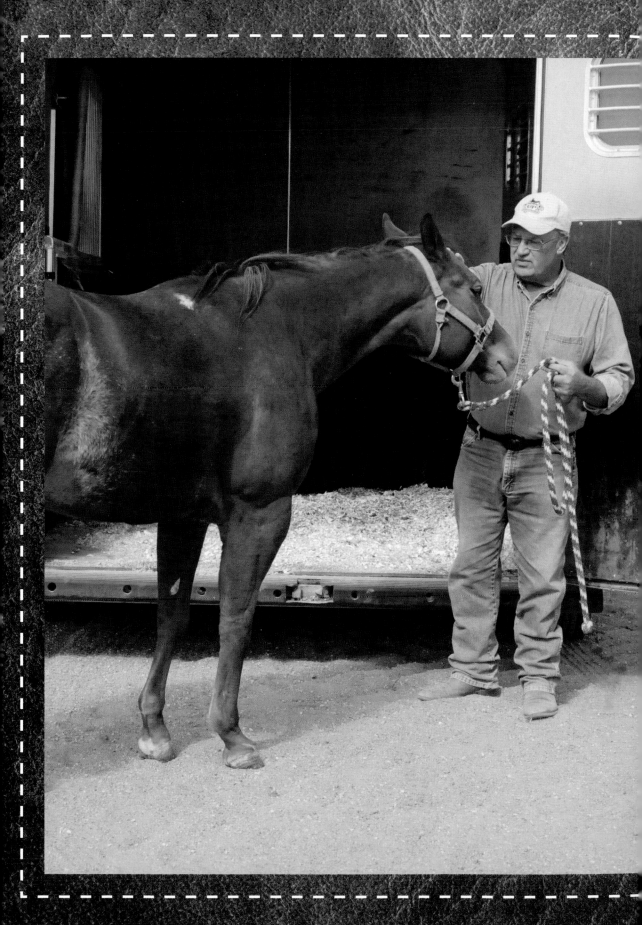

CHAPTER 5

TRAILER LOADING

Thanks for allowing me to share my philosophy about trailer loading with you. My techniques work on every breed and every type of horse. The key is to remain consistent. You will find that trailer loading takes a lot of patience; however, horses actually learn how to load into and back out of a trailer fairly quickly, especially if you are calm and gentle. Never get into a tug-of-war with the horse or you will lose.

Before teaching your horse to load, it is imperative to do roundpenning first to establish leadership. If you have no control over your horse, loading can get dangerous. This is especially true if there is an emergency and you need to get the horse in the trailer quickly.

Let me emphasize the importance of the size of the trailer. Beg, borrow, or do whatever it takes to get a trailer with the biggest and highest opening possible when starting your horse on loading (**Figure 1**). Later you can acclimate the horse to the more confined area of a smaller trailer, but it's best to always get the biggest trailer you can when starting out this lesson.

A lot of horses don't like getting in a trailer because it is so confining. As prey animals, horses view places like trailers as caves where predators like wolves would be and where they could be attacked. They instinctively want to avoid it. Also, because horses are gregarious herd animals, it is against their nature to be separated from the herd. Yet that is what we are asking them to do when we ask them to go in a trailer alone. It's little wonder we sometimes have trouble loading them.

I don't like struggling to get a horse in a trailer. Even if it were possible to force a horse into a trailer, I would never do it. I want the horse relaxed both going in and coming out. There are a lot of horses that will go in a trailer just fine but then come flying out backward. We want the horse steady and cooperative coming out as well as going in, so both should be done slowly and calmly.

Trailer loading can be taught to a horse just like any other lesson.
Nick Vedros, MindFire Communications

Figure 1. The ideal trailer should be big enough to easily accommodate your horse.

Figure 2. It is best to have friends on hand to help load a horse.

Because horses are gregarious herd animals, it is against their nature to be separated from the herd. Yet that is what we are asking them to do when we ask them to go in a trailer alone. It's little wonder we sometimes have trouble loading them.

For this reason and for your safety as well as the horse's security, make sure you don't take this lesson too fast. It doesn't work if you hurry to teach your horse to load too quickly. We want the horse to relax and trust us, so take your time.

Also, if you have the opportunity, it's best to have a partner on hand (**Figure 2**). This is recommended not only for your own security, but it's more fun. You can help teach each other as well as teach your horses.

When doing trailer loading or almost any training with horses, I love using the reward system, which is giving treats. I use the same grain horses eat every day to reward the horse when it does what I ask or if it at least tries. Using treats as a reward reinforces the lesson and is very effective.

There are several tricks that can be used to help load horses, including putting the horse's own manure in the trailer. Horses smell it and since it's familiar, they relax. A lot of people have another horse in the trailer to encourage their horse to load. I prefer to use the reward system and try to make loading a pleasant experience with or without the buddy.

For this lesson, we need to have the nylon string handy to use for applying pressure and release of pressure. This is the same type of string used in Chapter 2: Control Techniques. The horse will feel the pressure if you remain consistent, but don't use it for a tug-of-war or you will lose. If it's used correctly, the horse will learn how to release its pressure and when that happens, the horse will follow you into the trailer less reluctantly each time. Then treat the horse if it will take it.

The ultimate goal is to have your horse step into the trailer with just a verbal command. It's quick, easy, and stress free. Always quit on a good note, stay safe, and have fun.

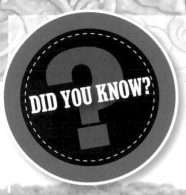

THESE GIRLS enjoy bathing their United Kingdom Shetland, a breed possessing an innate ability to connect with children of all ages. Though the ponies take their job with specially abled children seriously, they also know how to have fun.
Vanessa Wright, Personal Ponies

ROMAN RIDING on horses at the Fort Worth, Texas, Wild West Show requires trust between horses and rider as well as precision and courage.

David Roth, Austin Anderson, Texas White Horse Ranch

STEP-BY-STEP

Figure 1. The trailer is ready when doors are held open and inside panels are closed against the side.

Figure 2. I give the horse grain from inside the trailer because she stepped closer.

Figure 3. I lead the horse in a circle by the trailer before asking her to load.

Figure 4. I want the horse to stop when she is half in and half out of the trailer.

Figure 5. The mare backs out before I ask her to, and I allow it.

STEP-BY-STEP

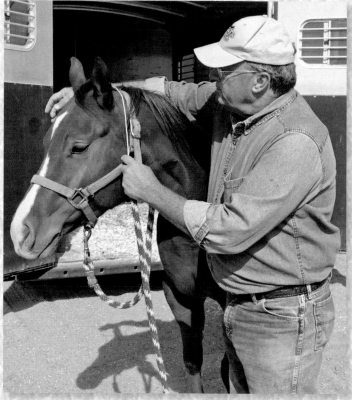

Figure 6. I hold one end of the string by the top halter ring and bring it over her head to the other side.

OVERVIEW

Begin trailer loading with a bucket of grain set just inside the trailer doorway to use for the reward system. Also, have a nylon string handy to use if necessary.

It's important to have the trailer door (or doors) braced open so it doesn't accidentally swing shut on the horse while you work with it. If there is a side door at the front of the trailer, it should be braced open also. It will make the horse feel less confined and can function as a quick escape for you if you need it. Make sure there are no sharp points on the trailer to hurt the horse and everything loose should be tied up (**Figure 1**). If you choose to put the horse's own manure in the trailer so it can smell it, now is the time to do so; this will make the trailer less intimidating. Also, get your partner on hand to assist with loading, especially if you are a beginning horse owner and are training the horse yourself.

Figure 7. I measure the string to make sure it is long enough by bringing it past the horse's lips to the left corner of her mouth on my side.

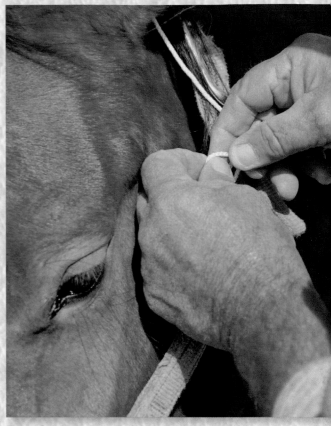

Figure 8. I tie the string to the upper side halter ring.

THE LESSON, PART 1

I am going to describe what this procedure looks like using a mare that is uncomfortable with loading. With a simple halter on her and a leadrope, I walk her up to the opened trailer. I let her relax wherever she stops because that's where her comfort level ends. I try to get her head down before asking her to come closer. She steps forward when she sees the grain, so I give her a little when she puts her head down for it. I introduce her to the inside of the trailer

when she steps close enough and let her sniff it. I step inside and give her another reward from there (**Figure 2**).

I want the mare to step up into the trailer with only her two front feet. (If there is a ramp on the trailer, you want only your horse's front feet on the ramp at this point.) I tell her "Up" and if she refuses to enter, I repeat the word, kiss to her, turn her head to the side to disengage her rear quarters, and put a little pressure on the leadrope. If she still refuses to get

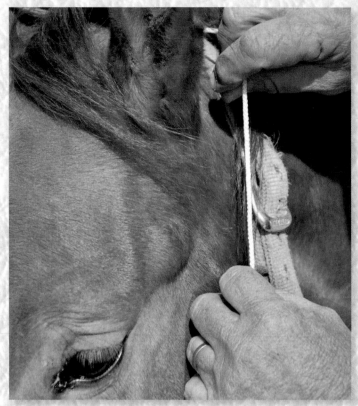

Figure 9. The string is brought over the horse's poll and checked to make sure it isn't on top of the halter.

in the trailer, I will ask her one more time just to give her a chance.

If she won't step into the trailer, then I step off the trailer and lead her in a circle (**Figure 3**) and up to the trailer again. If she steps into the trailer with her front feet, I tell her to relax there and reward her with a little grain. If she backs out, it's okay for now, and I still give her a little grain for trying.

I lead her around in a circle again and lead her up to the trailer. I only want her to put her front feet in the trailer to begin with when I say, "Up." If she steps in with her front feet, I let her stop there and relax while her rear is still outside the trailer and give her a reward.

Before getting her all the way in, I practice backing her: I say, "Back," and back her out, which should be easy with a horse that doesn't want to load. It is imperative that she backs out on command because we want horses to back off the trailer calmly. Then I lead her away from the trailer. If you have a

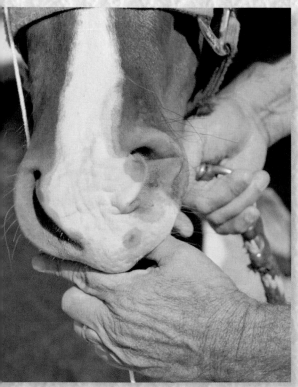

Figure 10. I bring the string down the other side of her head and insert my thumb under the front of her lip to lift it so I can run the string along her gums.

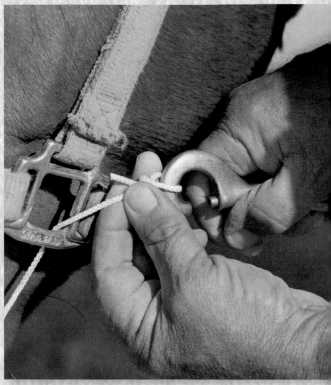

Figure 11. The string goes through the lower side halter ring and the leadrope is snapped onto the loop at the end.

horse that really likes the grain, you may have to pull a little to lead it away.

I lead her in a couple of circles and then back to the trailer, where I tell her "Up" again. If she steps into the trailer with her front feet inside (**Figure 4**), I give her time to investigate the inside of it, tell her to relax, and give her some grain.

If you get your horse to load partway like this mare did, let it take its time investigating the trailer and smelling it because it's an all-new environment. There may be smells from other horses in the trailer that it will want to know.

Be sure that you are not anxious when your horse gets this close to being loaded. That anxiety will project straight to your horse and make it anxious too. Both you and the horse should be comfortable with it half loaded like this.

I tell the mare "Back" and back her out of the trailer but ask her to load again right

Figure 12. The string fits closely along the top of the horse's gums.

away. If she steps in, I stop her with only her front feet in. When she relaxes, I give her a treat. It's important to wait until your horse gets its head down and relaxed before you treat it there.

Now I want the horse to step all the way in. I tell her "Up" and kiss to her to move in. If she backs out instead, she doesn't get a treat, but I will try one more time before moving to the next step. If she steps all the way in, I want her to stay there and relax. If she comes in but starts to back out right away, I tell her "Back" and allow her to back out because I want her to think it was my idea (**Figure 5**). It wasn't exactly what I wanted, but I still give her a reward because she got in when I asked her. I lead her around before approaching the trailer again.

If your horse happens to step up easily onto the trailer before you ask, back her out

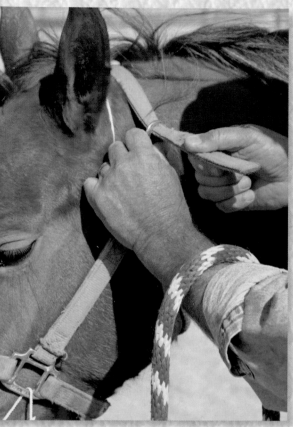

Figure 13. I tighten the halter so it won't slip when the string is pulled.

Figure 14. The horse steps into the trailer with a loose leadrope.

because you didn't ask her to do it. We want our horses to do it only on command, so turn your horse around in a circle again. The more time you spend with this, the better your horse will understand what you want and the more control you will have.

If my horse pauses the next time we approach the trailer, I step in and say, "Up." If she steps in with just her front feet, she gets a reward there and I tell her to relax. Then I say,

"Up," again and if she steps all the way in, she gets another reward and a chance to relax.

If she backs out on her own without my command, however, I allow her to do it, but I don't reward her. I won't fight with her if she wants to get out. I try one more time and lead her around, walk her up to the trailer, and say, "Up," which means putting her front feet in right away. If she does it, I tell her to relax and then say, "Up," again to get her all the way in.

If she's hesitating, I'll move the horse's head around a bit to help disengage her rear so she will move forward.

These are the steps:

1. Try to load the horse's front end only. Half a dozen tries are sufficient enough to give the horse a chance to enter the trailer.

2. If you get the horse's front end loaded, reward her, let her relax, then back her out. Do this at least two times before going to Step 3.

3. Try a couple of times to load the horse completely after it has loaded halfway.

THE LESSON, PART 2

If the horse refuses to load following the steps on Part 1, it's time to move on to the string technique because I want to make this as smooth and easy as possible. I use a type of string that's made of pure nylon so it won't break. (A roll of it can be purchased at any hardware store.) I cut it to about a yard and a half long, make a small loop on one end, and keep it in my pocket to use when I want to apply pressure on a horse. The horse learns to release the pressure itself by not pulling back. Normally this technique has a calming effect on horses and helps with an attitude adjustment if they need it.

First, we have to make sure the string is long enough to fit the horse's head. Standing on the left side of her head, I hold one end of the string up to the top side ring of her halter and bring the other end of the string with the loop on it over the top of her head to her other side and let it hang down the length of her head on that side (**Figure 6**). Then I bring it over her lips to the corner of her mouth on my side (**Figure 7**). I want enough length to reach the lower side halter ring and also have a few extra inches of string hanging down,

If you're ever in a hurry to do something with horses, don't do it!

including the loop I made previously. If there is an excess of string, I cut it off.

I tie the first end of the string to the upper side ring of her halter on my side of the horse (**Figure 8**).

I make sure the string is lying over the top of the horse's head and not on the halter (**Figure 9**) so it isn't touching anything except her, and then I run it down the opposite side of her head. I bring that end up to her mouth, and with my thumb lifting her upper lip, I run the string along the top of her gums (**Figure 10**). Then I string it snugly through the lower halter ring on my side of the horse. I don't tie it there, but snap the leadrope on the loop at the end of it (**Figure 11**). The weight of the snap alone will keep the string taut against her gums (**Figure 12**), so it won't slip off. Most horses mouth it a little and lift their lips, but soon adjust to it.

I tighten the halter (**Figure 13**) because it is imperative that it fits snugly in case it is dragged around the horse's head when she pulls on the string. It should be as tight as possible without being uncomfortable.

With the string in place, I walk the horse around again and lead her up to the trailer. I stop her before she steps into the trailer because I only want her to do that on my command. I say, "Up," and if she steps in with her front feet and stops, I tell her to relax, then say, "Up," again and kiss to her to come all the way in.

If she hesitates, I apply a little tension on the leadrope, which causes the string to put pressure on her gums as well as her poll. She might move backward, but it's important not to get into a tug-of-war with her. If she pulls back and backs out of the trailer, I let her, but I keep a consistent pressure on the rope and string.

It's important to use this technique wisely. The string is not a form of punishment; it's a gentle tool and should be used to encourage the horse forward. Never jerk on it to discipline her for backing up.

If she backs away, I kiss to her and say, "Up," again. If she steps forward, I quit the tension on the string, which releases the pressure on her gums. It also helps if I direct her head to the side with the rope. If she steps in with her front hooves, I stop her before she comes in too far. After pausing briefly, I say, "Up," and apply light pressure again. If she tries to dislodge the string by throwing her head up and backing out, I don't release the pressure. Usually horses step in again right away to release the pressure. If she steps in with all four feet, I reward her with a bit of grain.

Sometimes horses won't accept grain when they have the string in their mouths, but it should still be offered. If your horse accepts the grain, it will help reinforce the correct response. I don't know exactly why some horses refuse the treat, but I think they shut down and maybe pout a little. Either way, keep offering it. Don't offer a treat if the horse did the wrong response, like stepping in or backing out when you didn't want it.

Next, I want the horse to back up partway, until only her rear feet are out. I kiss to her and say, "Back," until she backs out. I stop her from backing all the way out; her front feet should still be in the trailer. It's important that there is no pressure on the leadrope here.

Take all the time in the world at this point. We want to keep the horse from flying backward as much as we want to keep her from lunging in.

After a pause, I tell her "Back" and she should back out easily. I lead her around and up to the trailer again. I only want her to load on my command and do it with a loose leadrope (**Figure 14**). I say, "Up," and if she steps up into the trailer with her front feet, I let her relax there. Then I say, "Up," again and if she backs out instead of coming all the way in, I repeat, "Up," until she steps into the trailer with her front feet. If she is half in the trailer, I say, "Up," until she comes all the way in, then give her a treat, and let her stop to relax.

I only use light pressure on the leadrope with the string on the horse's gums, and it doesn't take long for her to learn not to pull on it to keep the pressure off. When she doesn't pull back, there is no pressure on her mouth.

If the horse backs out on her own, I keep pressure on the rope and say, "Up," until she loads again. I tell her to relax when she is all the way in and reward her.

Then I say, "Back," and stop her when she has her hind feet on the ground. If her head is down and relaxed, I give her a treat inside the trailer (**Figure 15**). By stopping her halfway out, it teaches her to back out slowly and stay calm. Then I back her all the way off the trailer. I lead her away from it.

I have found that horses aren't on a time system; they live in the here and now. When I turn a horse away from a trailer, the trailer is out of sight, out of mind. When we head back to it, it will be like a new day to her. Because I didn't use force, she will have a new mindset and attitude toward it.

STEP-BY-STEP

Figure 15. I treat the horse when her front is in the trailer and her hind feet are on the ground. Her head is down and relaxed.

Figure 16. The horse steps completely inside the trailer on my command. The leadrope is very loose.

Figure 17. I remove the string on the halter.

Figure 18. The horse is now loading with just the leadrope around her neck and on my command.

As I mentioned before, the size of the trailer is so important. Some horse people truly love their horses, but they want to use their own trailers to train for loading, even if they are too small. This is where I always stress to beg, borrow, or buy the widest, tallest trailer to practice loading your horse.

I had an interesting experience once with a lady whose horse I fixed at a clinic. She had another horse at home that wouldn't trailer load and she had a small two-horse straight load trailer with the dividing panel between the two halves. I went out to see the horse and told her that her problem was that the trailer was too small: Her horse physically wouldn't fit in the trailer. The horse knew it, but she didn't.

Get the biggest trailer you can to practice with. When your horse is accustomed to that, then go down in trailer size if you have to (as long as the horse can fit).

I lead my horse again up to the trailer. She steps up onto it with her front feet without any pulling from me. She stops correctly there until I say, "Up," with a light tug and she steps fully on (**Figure 16**). Then I reward her.

It's the same way backing out. I lightly direct with the rope and say, "Back," and she steps back until her hind feet are off and I stop her there. I tell her to relax because her head popped up and I don't want her to rear. She puts it down on command (see Head Down lesson, pages 59–65), and I give her a treat. Only when the horse puts her head down will I give a reward. She is calmly standing half in and half out of the trailer, and I want her to be comfortable there.

Then I say, "Back," using a loose leadrope, and she backs off the trailer. I will load her one more time before I take the string off, but I will keep it handy either in my pocket or tied up around her halter out of the way in case I quit this part too soon.

I want the horse very relaxed and loading only on my command. I walk her around and step into the trailer. When I say, "Up," she steps in with her front feet only. I stop her there and tell her to relax, then again say, "Up." She steps all the way in. It's exactly how I want her to load. She is relaxed and looking for her treat, which is cool, so I reward her.

It's the same way backing out: I want her calm and focused (although I will let her move if she wants to get at flies), but I don't want to be in a hurry. If you're ever in a hurry to do something with horses, don't do it!

I say, "Back," and she steps off with her rear feet only. Her head is down and relaxed and I give her the treat as she stands half off the trailer. Then I say, "Back," and back her totally off.

THE LESSON, PART 3

Now it's time to remove the string because the horse did the lesson well two times in a row (**Figure 17**). I lead her around with a loose leadrope and trust that she will load without any tension on it.

When we approach the trailer, I say, "Up," and the horse steps in right away with her front feet and the leadrope very loose. I tell her to relax and then say, "Up," again and she steps all the way in. I give her the treat. She is very relaxed with her head down. We still have to back out, so I say, "Back," and kiss to her with light pressure on the rope. She backs off halfway and takes her treat there while she is half on and half off. She is calm, and when I tell her again to back, she backs out.

Now I throw the leadrope up around her neck and lead her around in a circle. When we return to the trailer, I just use the verbal command "Up." She steps easily into the trailer the way I like it. There is a pleasant reward each time she loads properly and I don't rush her. I move her farther forward inside the trailer because I want her at the front. She relaxes there and gets treated, then I say, "Back," and she backs halfway off. I tell her to relax and say, "Head down" (which she learned before). She drops her head. Then I say, "Back," and she backs totally off the trailer.

I lead her in a circle and up to the trailer again. I throw the leadrope up around her neck and stand beside the trailer this time. I say, "Up," and she doesn't hesitate to step inside (**Figure 18**). She moves all the way to the front of the trailer to get her grain in the bucket, and I let her enjoy her time in there and relax with her treat.

It's good to also desensitize her there by knocking on the side of the trailer to make noise as long as she is relaxed. Then I kiss to her and say, "Back," and without touching her, she backs halfway off before I direct her to come all the way out. I load the horse one more time starting at a greater distance from the trailer. I throw the leadrope around her neck, which will be her cue to trailer load from now on. I stand way to the side, point her at the trailer, and say, "Up." She doesn't hesitate to walk up to it and go into it all the way to the front.

She's doing so well that we are going to quit here on a positive note, let her enjoy the grain, and let her be calm, cool, and collected. We want this to end with her remembering it as a good experience.

SUMMARY

It's important to teach a horse how to enter and back out of a trailer calmly and on your command. Don't think that it's okay for your horse to turn around in the trailer to come out. It should learn to walk directly in and back directly out.

There may not be enough room to turn a horse around once it is loaded in the trailer, and the horse should know how to back out. It has to trust you and come out calmly. You have to remain consistent and do it properly every time.

I want those back two feet to hit the ground and stop. If we get away from being consistent ourselves, we can't ask the horse to be consistent.

The last thing we want to do is to traumatize a horse while loading. The horse won't easily forget it and there will be a battle the next time the horse needs to be loaded. With this lesson, there is no force used. If our horse trusts us as the leader, it will always try what we are asking it to do.

The mare munches away at her treat in the front of the trailer and she's happy.

> **Total time for this session: 17 minutes**

Dennis directed these exercises:
1. Taught the horse to calmly load halfway in the trailer using only a verbal command.
2. Taught the horse to step completely into the trailer and move all the way to the front of it on his verbal command.
3. Taught the horse to calmly back halfway out on his verbal command and stop.
4. Taught the horse to back out all the way on his verbal command.
5. Taught the horse to load on the trailer on his verbal command without leading her and also to back out of the trailer by his verbal command while he stood outside the trailer.

THESE RIDERS are having a fun time on the beach with their Marsh Tacky horses at the annual Carolina Marsh Tacky races. Even at a full gallop and out in the open, their horses are under control.

Dwain Snyder

LEADERSHIP ISSUE/ NO CONTROL

A horse that doesn't recognize you as the leader is particularly difficult to load into a trailer. Try as you might, that horse won't cooperate when you ask it to go into the small, threatening space of a trailer. First you need to earn the horse's trust in the roundpen, then you can begin work on trailer loading.

An inexperienced owner brought a horse to me that has issues with loading. This is her first horse, a young mare, and the owner doesn't know how to control her or be a leader to her. The mare has never been successfully loaded in any trailer, not even large stock trailers. The owner had to ride her to get her to me and is hopeful I can train her horse to go in a trailer. I'm hopeful too.

First, I worked the mare in the roundpen for twenty minutes and established leadership and control to the point where it is possible to start on her trailer-loading issues. Then I put a simple halter and leadrope on her. The lead-rope is cotton, not nylon, which would burn my hands if I had to pull on it.

THE LESSON

I walk the horse up to the owner's open-stock trailer that we will be using. Its doors are braced open so there is no chance they will swing closed on us during loading.

There is a bucket of grain handy by the trailer, and when the horse approaches it, I let her have a taste. I step into the trailer and kiss to her to follow, but she refuses to go in. I walk her around and try again, just to give her a chance. I move her head to the side to disengage her rear quarters and she steps forward but off to the side of the trailer to avoid getting

in. I give her a bit of grain as a reward because she did step closer to the trailer. I kiss for her to follow me in, but she backs away.

I never fight with a horse to load, so it's time to take out the nylon string I always keep in my pocket (**Figure 1**). (To put the string on the horse, see Trailer Loading lesson Part 2, page 194) I will use it for my safety as well as the safety of the horse. It will put pressure on her poll as well as her gums when she pulls back. It won't be painful, but it will irritate her and she will learn quickly how to relieve that by not pulling against it.

I lead the horse around and let her mouth the string to test it and get used to it. It will feel like dental floss stuck on her gums. If she pulls on it, it will be a little uncomfortable and will teach her not to resist, but if she didn't have it on and pulled hard during the loading process or reared up, she could hurt herself, and we don't want that.

I walk the horse up to the trailer, get in, and only apply a little pressure on the rope. The technique is not to use it as a tug-of-war with the horse. If she backs away, I will go with her without pulling or tugging on it. As soon as she relinquishes pulling, the pressure will quit and she will come forward on her own without any pulling from me.

The horse responds by raising her head and tries to pull back, then puts her head down and steps forward, releasing the pressure on her gums. I kiss to her and direct her head at an angle to the trailer and she steps up to it, but not in it. I put a little pressure on the rope, and she takes another step. She is trying to cooperate, but I notice that if she steps into the trailer, her head will almost touch the ceiling no matter how well she goes in. So we give up on this trailer and bring her to a bigger one.

It never pays to be in a hurry to teach a horse something. If it takes all day but she learns to load, the lesson will last the rest of her life.

This is so important. Don't use a trailer that is too small or too confining for your horse. Horses are prey animals and don't have the same reactions to small spaces as us predators. They are not used to being in confined places like we are. Always get the biggest trailer you can get, and you'll find that loading your horse will go much easier.

When I lead the mare to the bigger trailer, she stops a few feet away. She still has the string in her mouth. When she leans back, I lean with her and don't pull any harder (**Figure 2**). If she steps back, I will move with her because I don't want to cause pain for either one of us.

The horse pulls back and rears up, which doesn't affect me at all. She is safe and so am I.

I walk her around in a circle to calm her and let her head come down in a more relaxed position. She balks at approaching the trailer again. I kiss to her and put a little pressure on the rope, but she raises her head and steps back. She is putting pressure on her gums by pulling back. She brings her nose up, trying to dislodge the string, but she doesn't really fight it because it is just irritating, not painful.

We have a standoff there. She sticks her nose all the way up and I allow that without adding any additional pressure on the rope. I don't pull her, and she finally gives up on her own and steps forward.

If she really pulled back, I would have done my best to retain the same amount of pressure, which isn't painful for her, just uncomfortable. She is applying pressure to herself when she pulls, but she will learn soon to give to it. There is nothing violent, aggressive, or painful with any part of this. The horse isn't struggling to get free because of fear, intimidation, or pain. There is no fighting or struggling. We are doing this slowly at the horse's own pace and letting it be her decision when to step in the trailer. I always let a horse have all the time in the world to decide to enter a trailer because horses can't be rushed.

Finally she steps close enough for her head to be inside the trailer. I get her head down and relaxed by giving her some grain as a reward. She checks out the inside of the trailer, and I allow plenty of time for her to look and smell around. It never pays to be in a hurry to teach a horse something. If it takes all day but she learns to load, the lesson will last the rest of her life.

A little trick I do sometimes before a horse checks out a trailer is to put its own manure inside. The smell of her own manure in the trailer will make the horse more comfortable with loading, but we don't have any by this trailer.

I kiss for her to step inside, but she raises her nose and tries to struggle against it. I won't

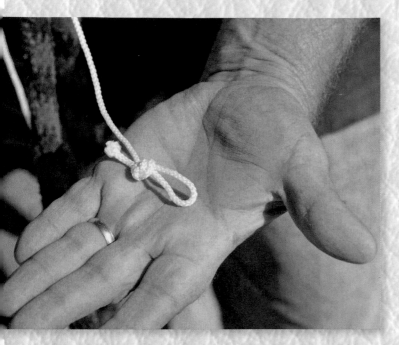

Figure 1. This is the nylon string with the loop at the end that I tied there.

Figure 2. The horse stops short of the trailer and I keep the rope tension light.

Figure 3. Finally, the horse decides to step inside.

STEP-BY-STEP

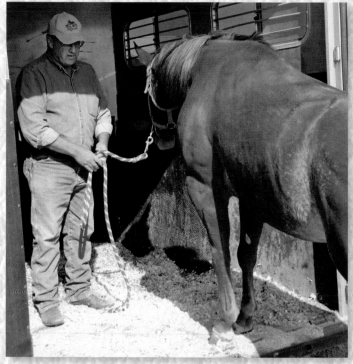

Figure 4. The horse pauses inside the trailer and I let her look around.

fight with her, so I follow as she steps back without applying any additional pressure. When she brings her nose down and steps forward, I release the small pressure I had on the rope and she relaxes.

I step in the trailer again and this time she follows and puts one hoof tentatively on the floor (**Figure 3**). All right! I let her pause there, and soon she steps up into it with her other front hoof. She is standing half in and half out, and I give her a treat as she pauses there. She isn't anxious or fearful. I readjust the halter so the string is correctly placed on her poll, not the halter. She looks around and

then backs out. I follow her out and walk her around. Progress!

When I lead her back to the trailer, she steps up to it quickly this time. She tries to pull back once and then gives in and comes close enough that her head is inside. We have a standoff there again, but I am consistent with a little pressure on the rope and she tries to move her head to avoid it. She feels tentatively with one hoof on the trailer again, then pulls back. I quietly wait for her inside the trailer while keeping a steady (but light) pressure on the rope. She knows that pulling on it causes pressure on her mouth, and she knows how to relieve it now.

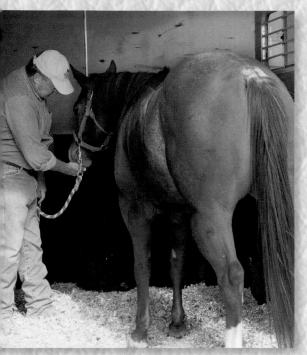

Figure 5. The horse comes all the way in the trailer and I give her grain.

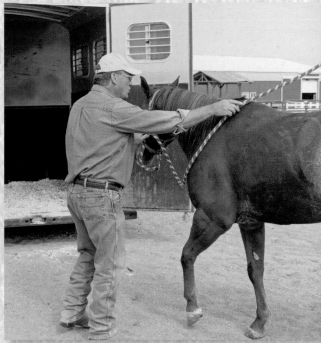

Figure 6. I throw the leadrope over the horse's neck and she loads herself in the trailer.

The horse pauses and is quiet while she evaluates the situation. She isn't acting afraid, nervous, or apprehensive. All her objections have been taken away: This trailer is big enough to accommodate her, and there hasn't been any pain associated with it. I am only politely asking her to get in.

She pulls back one more time, trying to dislodge the string, but I kiss to her and she steps into the trailer (**Figure 4**) with both front feet. I stop her there and let her relax. I adjust the string on her poll and kiss for her to come all the way in. She pauses there for a while, and I wait patiently. Then she lunges backward and backs out quickly, so I follow her out.

Some people lose their tempers or get excited at times like this when the horse is so close to loading and backs out instead, but it is never an option to get angry or be in a hurry. The horse did step partly into the trailer when I asked, and she is trying. She has never been in a trailer before and now has to learn how to step up into it.

I am having a good time and joke with the owner who is standing nearby. We are happy the horse is learning to load, and we aren't anxious about it at all. I walk the horse away from the trailer and let her relax because I don't want her upset about backing out.

When I bring her up to the trailer and get in, she balks a little at the entrance, then gives to the string and steps up closer. There is a bit of a standoff again as she plants her feet, but I keep the pressure consistent with the rope on the string. She moves around, puts one foot in, takes it out, and eats the loose hay on the trailer floor. I don't want her rewarded there so I lead her away in a circle and then back again to the trailer. As I get in, she tries to follow and puts one foot in, then the other, and pauses there like she did before.

I put a little pressure on the rope and try to get her in all the way by moving her head around a little. I want to give her all the time she needs and help her make the right decision by applying a small amount of pressure with the rope. I kiss to her and encourage her, but she backs out, then steps back in right away with her front feet.

She pauses a long time, then steps fully inside and all the way to the front of the trailer. Cool! That's her first time in a trailer. She gets rewarded with grain there while the owner whoops with joy beside the trailer.

Once a horse learns to load, it's good to desensitize it to the noises a trailer can make while it is inside because a horse can get nervous with that, so I move the partitions a little. I encourage the owner to work on doing lots of desensitizing with her horse to trailer noises when she gets home.

Because this is the first time the mare has ever been in a trailer and she already learned a lot today, I turn her around and walk her out instead of making her back out. I walk her around and lead her back to the trailer, saying, "Up," as I get into it, and she steps in

Trailer loading should be as positive an experience for horses as possible.

right away. I tell her to relax, reward her, and lead her out again. The owner is ecstatic.

I walk the horse around and lead her back in. She doesn't hesitate to step in the trailer when I say, "Up." I stop her when she is halfway in and tell her to relax. Then I say, "Up," again and she steps in all the way without a problem. I let her relax and reward her there with grain (**Figure 5**).

I repeat the process and have the horse pause when she is halfway in before letting her move all the way in. She is cooperative and loads easily now. After I take her out of the trailer, I go ahead and remove the string from her gums. Sometimes I make a mistake and remove it too early, before the horse has really learned the lesson, but I keep it handy in my pocket and can always put it back on.

I walk the horse up to the trailer, and she loads with no problem. Now we switch from myself to the owner, who comes in the trailer to reward her horse while I leave. The owner leads her horse out, turns her around, and leads her back in the trailer easily. When she takes her out the next time to lead her around, I coach her to pause by the side of the trailer and tell the horse "Up." When she does, the mare steps easily inside and gets a reward.

The next time the owner brings her out, I show how to throw the leadrope on the horse's neck (**Figure 6**), point her at the trailer, stand to the side, say, "Trailer," and she steps immediately into it. These will be her cues to load from now on.

I bring the horse out, lead her twenty feet away from the trailer, point her toward it, throw the rope over her neck, say, "Trailer," and she walks straight up and into it. The owner is very excited.

SUMMARY

I advise the owner to buy a trailer that has plenty of height to it. The partitions should be padded and the flooring should be solid, but it doesn't have to be brand-new. It can be a twenty-year-old trailer and still be in good shape and work for her as long as it accommodates the height of the horse. The trailer that we tried to use before was a foot too low in height, which makes a big difference when loading a horse.

If this lesson were done in a secure area where there was fencing all around, we could practice bringing the horse back farther away from the trailer, and if we were consistent, we could point her at it from any distance, say, "Trailer," and she would learn to load herself from there. It's fun to see how far we can get away from the trailer to do it.

We let the horse relax and eat her grain in the trailer. Trailer loading should be as positive an experience for horses as possible, and it is positive when they can enjoy some grain in the trailer. It's not bribery; it's to make them feel comfortable there. People are the same way: If we do something that is comfortable for us, we want to go back to it.

The owner practices loading her horse several more times with just the rope thrown over the mare's neck and standing twenty feet from the trailer. The horse walks straight up to the trailer and loads herself every time. Cool! The owner is very excited that her horse will not only load for her now, but that she will load so easily and at this distance.

> **Total time for this session: 29 minutes**

Dennis directed these exercises:
1. Taught the horse to come halfway in the trailer and pause so she learns to walk in calmly on a verbal command.
2. Taught the horse to walk all the way into the trailer on command.
3. Taught the horse to walk into the trailer by herself and from various distances using only a verbal command and without anyone leading or guiding her or using any other aids.

WON'T BACK OUT OF TRAILER

Some horses load into a trailer just fine, but they don't like backing out. These horses like to turn around inside the trailer before they come out. That might seem like a reasonable solution, but for several reasons it's not.

Number one: Sometimes it's not possible. In a smaller trailer, a horse can't physically turn around. I know what it is like to have this problem. Many years ago, I was giving a training clinic in Colorado and had a really skinny trailer to load a big horse in. Guess what I forgot to do? I forgot to teach him to back. I was pinned in there after I loaded him. Thank goodness no one got hurt and I eventually got him out.

Even if you own a really big trailer that your horse can comfortably turn around in, you never know when it may have to be in another trailer that's smaller. Just like with everything else in training horses, consider the future and all scenarios.

Number two: It's unsafe for the horse, especially in a slant trailer. In this kind of trailer, if the horse slips or gets bumped while turning around, it could end up with hip injuries. The horse could even fall, which would be serious.

Today I am working with a gelding that is a great trail-riding horse, and the only issue the owner has with him is that he won't back out of a trailer. His original owners had a huge stock trailer. There was no problem in letting him turn around inside the trailer before he came out, but it's a big issue for the current owner.

His current owner has a large three-horse slant trailer. It's large enough for her to physically turn the horse around, but he has to go in the middle part of the trailer where there is enough space to do it. He can't be the last horse in because there isn't enough room for turning around in that area. Also, it's a slant trailer, making it more dangerous.

There have been many attempts to back the horse out of a trailer, but he pushed into the person unloading him and forcibly turned his head and body around. The owner worked with him for a couple of weeks, trying to slowly back him out, but to no avail. She tried to get his front feet in and back him out from there, but he refused. When her husband tried, within thirty seconds the horse got mad and bit him. The owner spent hours working with the gelding one day and was determined to get him to back out, but ended giving up, so the horse won again.

Horses are extremely powerful animals and can't be forced into anything. Anyone who tries to make a horse do something it doesn't want to will soon find that out, as this owner did. Her last hope is that I can train her horse to back out and also teach her how to do it. It would end the hassle with loading her horses for road trips and would be safer and more pleasant, both for the horse and for the people unloading him.

The gelding also tried to bite me and definitely needed an attitude adjustment, so I worked him in the roundpen first to establish leadership before confronting the backing problem. I always want to control movement and establish leadership in the roundpen first, no matter what the issue is.

THE LESSON, PART 1

I begin fixing the problem by using the string technique: I put the string on the horse (**Figure 1**) and make sure the halter is on snugly. (See Trailer Loading lesson, pages 194–195.)

Figure 1. The horse adjusts to the string on his head.

The string will feel to the horse like what a little string on our gums would feel like: It doesn't hurt, but it makes the horse pay attention. The gelding reacts by stretching out his neck, lifting his lip, licking his mouth, sticking out his tongue, and mouthing the string. Most horses do this but usually adjust to it soon enough.

The owner stands by the trailer door, which is braced open so it won't accidentally shut on us during the training. It's a large three-stall trailer so we'll have plenty of room to work. There is a bucket of grain just inside the trailer entrance to use for the reward system.

I lead the horse up to the trailer and offer him a small handful of grain from the bucket. Some horses won't take grain with the string

in their mouth, but he accepts the grain without hesitation. Great!

The horse has to back up well for me before I can load him, so I try to back him by putting light pressure on the rope (and consequently the string), bringing it toward his chest. I say, "Back," and offer him grain with the other hand under his head so he will want to back up. Instead he swings his head at me in protest, so I block him with my arm, but he does back up a couple of steps and I let him have the treat.

I repeat the process of leading the horse forward and backing him away from the trailer four more times, each time asking him to back more steps and rewarding him with a bit of grain. This teaches him to back with little pressure on the rope, and he starts backing well without swinging his head toward me.

Since the owner is close enough to help, I explain to her what I will do next: I am going to make the horse step into the trailer with only his front feet when I tell him "Up" and stop him there, then make him back out. This will teach him to load and unload slowly and calmly, not rushing in or out. I will do this a few times before getting all four of his feet on the trailer. When that happens, I want his back two feet as close to the edge of the trailer as possible because the closer I get him to the edge, the easier it will be to get him off. Because it's such a large trailer, there's a lot of space for him to come forward, but I want him to stay back by the edge for now. Normally horses will shake and quiver when their hind legs are at the edge of the trailer, and I will only reward him when he quits shaking.

When the horse does back out, the owner and everyone else has to be out of the way in case he comes out fast, or comes out at the side instead of down the middle of the trailer, or rears up. (Family members are watching a short distance away and need to move quickly if he backs out fast and doesn't stop.) He could go up or down, sideways, or any way he wants. This is all new to him, and we don't know what his reaction will be.

I take a handful of grain and move the grain bucket farther into the trailer to have it handy when I go in there. I lead the horse closer to the open end of the trailer and make him back away from it five more times using the lightest pressure possible on the leadrope and rewarding him each time he backs up. I make sure he backs four or five steps easily each time and when I say, "Back," I raise my voice a little. It's stern, but it isn't yelling at him. (That's why no one calls me a horse whisperer!)

It is imperative I do this preliminary work here so I don't get hurt trying to make the horse back up when we are in the trailer. He must understand what "Back" means and do it with barely any pressure on the rope. He backs up quickly now with only a light touch from me on the rope. I'm satisfied with his response and am enjoying working with him.

I get in the trailer, and the horse walks up to it easily. I say, "Up," and kiss to him to follow me in. He steps in with his front feet right away and I say, "Whoa," to stop him there. I wait for him to relax and reward him with a bit of grain (**Figure 2**). Then I say, "Back," and he backs out easily with no pressure at all on the rope. I tell him to relax and hand him some grain. I repeat the same procedure one more time to make sure he is comfortable getting on with only his front feet and backing off the trailer without completely getting on it.

This is all about control: I only want the front two feet loaded and the horse to stop before he comes all the way in. It's so important because I don't want him nervous and running over the person loading him or anyone else who is around. I want him to come in and back out slowly and respectfully when I say so.

The next time I tell the horse "Up," he steps in with his front hooves and when I say, "Up," again, he steps all the way in. I am able to stop him with his hind hooves near the edge of the trailer floor and reward him there. I want him as close to the end of the trailer as possible.

Because his rear is positioned close to the edge, I assume his hind legs will quiver when I ask him to back out. I will correct him a little only if I have to, but I won't jerk on him because it might cause him to panic and rear. Horses can't see their back feet or the end of

Figure 2. I stop the horse when he is half in and give him a treat.

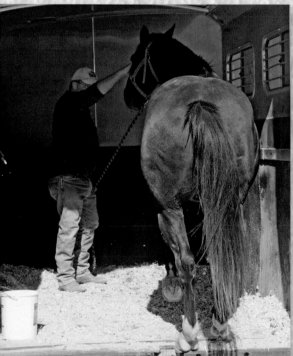

Figure 3. I put a hand on the horse's nose bridge to hold his head away from me while he takes a step back. He is close to the edge.

Figure 4. The horse is nervous with his feet so close to the edge of the trailer and he tenses up.

STEP-BY-STEP

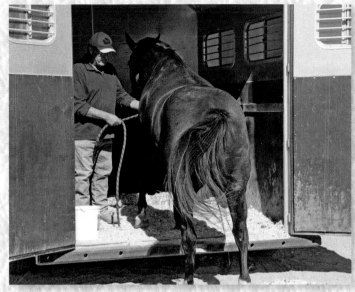

Figure 5. The horse backs halfway off the trailer and I stop him there.

the trailer when they are this close to it, so it can be scary for them.

I say, "Back," and kiss to the horse. He turns his head to see behind him and throws his head around. He angles himself away from the end of the trailer, trying to turn, but I won't allow it. He throws his head at me and tries to struggle, but he succeeds only in moving forward a bit because of the pressure on his mouth with the string.

I tell him to relax; I only want him to back up an inch at a time. I say, "Back," and put a hand on his nose bridge to keep his head in place. He takes a small step back (**Figure 3**). I let him stop and relax, then say, "Back," again and kiss to make him move. He jerks his head to the side but steps back, and I reward him with a small handful of grain.

Sometimes horses get upset with this procedure: They can get pouty and cranky and will even hold their breath or refuse grain, but all I want is one step back.

I tell the horse to back up again, and he shifts his feet around trying to comply without actually stepping off the trailer. When he takes the smallest step back, I stop him and reward him. This should be as positive an experience for him as possible. I let him relax and think about it while I adjust his halter. Teaching the lesson isn't something we need to rush. Anytime you're in a hurry to do something with horses, don't do it! It doesn't work!

The horse is at the edge of the trailer now. I ask him again to back up and he reaches down with one hind hoof and touches the ground, then steps back inside the trailer. I

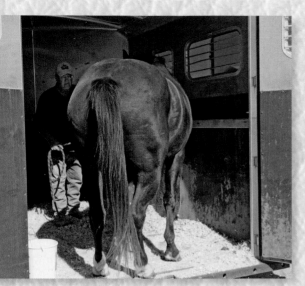

Figure 6. The horse stands with his rear at an angle in the trailer.

Figure 7. The horse backs out of the trailer at an angle while I try to keep his head straight.

tell him to relax and give him a reward. It was far from perfect, but it's one step more than we had before. When I ask him to step off the trailer again, he tries, but he is nervous and his rear muscles bunch up (**Figure 4**). He moves his feet around, trying to avoid the edge of the trailer, but does manage to move back just a bit more and we pause there. I can see his back leg quivering, and I want him to relax. If I push him too much now, it would give him an excuse to jump on me.

After the horse relaxes, I ask him to take another step back and kiss to him. When he gets his rear feet close to the edge, he starts quivering again, so I use only light pressure on the rope. If I were to push him too much here or jerk on him, he might go over backward. He struggles a bit but manages to back half

out by putting his hind legs on the ground. He tries to back totally off, but I say, "Whoa," and stop him half off the trailer. Then I reward him as he stands there.

This is the way I want him to back off the trailer: on command and only a little at a time. As soon as those hind two feet hit the ground, I want him to stop. Anyone could be standing close by when a horse is unloading, including kids. If that horse comes out fast, it would be especially dangerous, so I want horses to learn to come out of a trailer slowly.

I kiss to the horse and back him completely off the trailer, which he does easily. I reward him with a little grain from my hand and walk him around to let him calm down. Because horses have no concept of time, every time I take the horse away from the trailer for a short walk and then bring him back, it's like starting all over in a brand-new day for him.

I lead the horse back up to the trailer, and he follows easily. I get into the trailer and say, "Up." I have to put a little pressure on the rope to get him close to the trailer, but I release it right away when he does. He balks just a little, then steps into the trailer with his front feet. I stop him there, tell him to relax, and treat him with grain.

Then I say, "Up," again, and he steps into the trailer with his hind feet. I stop him just after his hind feet step onto the trailer floor, but his right hind leg is too far in and he is standing at an angle to me. Though it's not perfect, I give him a reward when his head is down. I want to adjust him, but his hindquarters are quivering, so I wait until he quits.

Then I say, "Back," and am able to get his back right hoof even with the other and closer to the edge. Great! All I wanted was that hoof a couple of inches back. If he refused, I wouldn't force him, but I got the two inches and rewarded him so he will think it's not so bad to back up.

I ask the horse to back up again, but he starts quivering and steps back at an angle. I try to keep his head straight so he can't turn around. He moves his feet a little, and I get him lined up against the trailer wall so he isn't standing at an angle. He steps around, and I have to straighten him again and tell him to relax. I give him some grain for trying, and he relaxes and quits quivering. If this were a situation where he was being unloaded and little kids were around, it could be a disaster because of how anxious he gets by just backing out.

I ask the horse to back, and he steps straight back a step or two to the edge of the trailer. I stop him there. I tell him to relax but don't give him a reward because his hindquarters are

I don't want to work the horse too much in his first lesson.

quivering a little, and he only gets rewarded when he doesn't quiver. When he relaxes, I ask him to back out. As I do, I'm careful not to use any tension on the rope. Though he throws his head around, he steps back until his hind feet are on the ground and I stop him there (**Figure 5**). I had to pull a little to prevent him from backing out with all fours at once. When his head comes down and he quits quivering, I give him a grain reward, then back him completely off the trailer and lead him around outside.

Always stop a horse when it is half off and reward it there if it puts its head down for a treat, which will help it relax. Make the experience as positive for the horse as possible. Every time a horse does the right thing, even if it's a little movement, I reward it. When I start colts or anything I do for the first time with horses, I over-reward and under-discipline.

I practice backing the horse away from the trailer entrance more and make sure the halter hasn't slipped and the string is on correctly. There is no pain or intimidation with this method, and it's obvious the string doesn't hurt him a bit.

I lead the horse forward, get into the trailer, and say, "Up," using light tension on the rope. He steps into the trailer right away with his front hooves and stops, so I reward him there. I ask him to come all the way into the trailer when I say, "Up," again, and I stop him as close to the end of the trailer as possible. I don't want to bring him clear up to the front of the trailer just yet.

When I ask him to back up closer to the edge of the trailer, he tries to turn at an angle, but I keep his head facing straight forward. He throws his head around but backs up a couple of steps, so I stop and reward him there. I check to make sure he is not quivering: It would ruin everything if he was quivering and I forced him to move back.

The horse is standing at an angle again (**Figure 6**), and I try to straighten him out and not bring him forward at the same time. Then I ask him to back up using only light pressure on the rope. He is trying but insists on moving his rear at an angle to the edge. He reaches down with one hind hoof, feeling for the ground, then puts it back in the trailer. I keep asking him to back, and he puts the hoof down again on the ground and steps down with the other hind one. I laugh because he acts like a ballerina touching the ground with a pointed toe.

The horse is standing half in and half out, exactly what I want, so I tell him to relax and reward him. I ask him to step back up on the trailer ("Up") and he does, so I reward him there. He is standing at an angle again, so I try to get him to stand straight by lining him up against the wall and then ask him to back. He takes a step back and starts quivering, so we pause and relax before I treat him. I want him to stay calm and only take a few steps back at a time so he doesn't get nervous and jump on top of me. I ask him to back again and he steps around, feeling with his feet until he steps off the trailer with his back feet and gets rewarded. I ask him to back again and he steps totally off the trailer and gets a big reward.

I'm not rushing because that wouldn't be a positive experience for him, and we need him to stay calm. He's not sure about that last step because it's a long way down off the trailer. I walk him around a little and make sure everything is adjusted on his head, then lead him back to the trailer. When I ask him the next time, he loads correctly with both his front and rear feet, so I reward him.

This lesson requires a lot of repetition. With light pressure on the rope, I ask him to back up, and when he takes a step back, I stop and reward him. I ask him to back again, and he moves back at an angle. I straighten him out, but he goes back to an angle before he steps out of the trailer, this time with all fours. (**Figure 7**). Nevertheless, he is unloading much easier now.

To give the horse a break, I walk him around, and when I ask him to load again using a loose leadrope, he easily steps in with his front feet and on command steps in with his hind feet. I reward him and check my string to make sure it is positioned correctly. When I say, "Back," he steps back and this time stays straight as he steps off the trailer with his hind legs. Nice! I reward him there and then back him completely off and walk him around. He backed off the trailer correctly this time and much quicker than before.

It's so much fun when I can just say, "Up," and the horse steps right away into the trailer with its front two feet, and when I say, "Up," again, it steps into the trailer with its back feet. It's the same when I can back a horse out of a trailer with no pressure on the rope and its head is low and relaxed. When a horse is like this and it isn't nervous or in pain, and there were no drugs used to calm it down, it makes trailering fun!

Some owners have to resort to drugging their horse to calm it down enough for trailer loading. But we can accomplish something

STEP-BY-STEP

Figure 8. I lead the horse to the front of the trailer and he relaxes there.

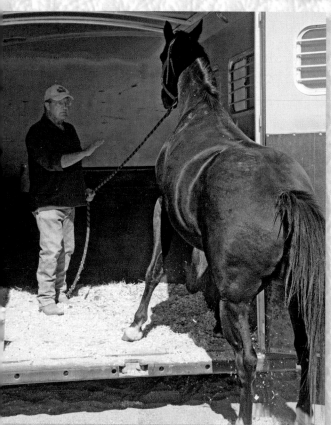

Figure 9. Before backing him off the trailer, I motion with my hand to bring his head down because it is too high.

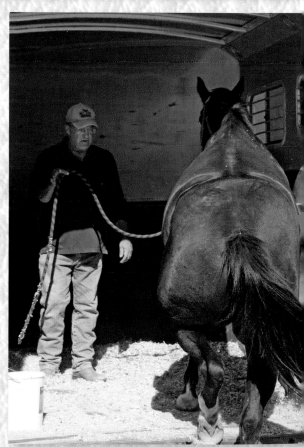

Figure 10. The rope is loose and he backs out fairly straight now with his head down.

much better through simple, consistent training. With my technique, the horse will calmly step into the trailer when you say, "Up," and calmly back out when you say, "Back." No drugs needed.

I repeat the process: When I say, "Up" using no pressure on the leadrope, the horse steps into the trailer right away, and when I say it again he comes completely in. He is coming in willingly, knowing he will be backed out, which is something he was afraid to do before.

Now I move him closer to the front of the trailer because he should get used to being there (**Figure 8**). I tell him to relax. His head is down and he looks comfortable. He backs up two steps for me and I reward him.

The next time I ask him to back, it's almost like he is tiptoeing with his hind feet as he feels around for the drop-off to the ground. I move him forward to straighten him out and ask him to back using a really loose leadrope. He backs up several steps at an angle again. I straighten him out, but he insists on standing at an angle, and the next time I ask him to back up, he steps back and to the side. I turn his head to straighten him out and back him several steps before stopping because he is getting nervous and quivering. He wants to turn in the same direction he used to do when he was allowed to turn around in the middle section of the trailer. I won't let him do that, but I won't back him too much when he is nervous.

I let him relax and reward him; then I get him to back up several more steps, and he successfully backs off the trailer with his hind feet. I stop him and use a hand motion to bring his head down (**Figure 9**); he follows my hand down with his head. I ask him to back completely off the trailer, and when he does, I reward him.

THE LESSON, PART 2

The horse is backing out well, so it's time to try the lesson without the string. I lead the horse away and remove the string from his mouth but keep it handy in my pocket in case I need to put it back on. When I bring him back to the trailer, he steps in correctly right away and I treat him after his hind legs clear the end. I ask him to back out without any pressure on the rope, just using the verbal command, and he backs until his hind feet are on the ground. We pause there and he gets a treat.

Then he steps off before I tell him to, but I say, "Back," anyway. When a horse makes a move like that without the command, I tell it to them as they are doing it. It will help them relate the verbal command to the action. So if you see a horse make a movement that looks like it is going to step up into the trailer, say, "Up," even though the horse is already doing it, or "Back" as the horse is doing it.

The owner is amazed that her horse is backing out so easily. Before I turn over control to her, I want to back the horse out of the trailer a few more times. I get him loaded and bring him to the front of the trailer by the first stall. I let him think about it, then ask him to back with no pressure on the rope; it actually has a dip in it. He backs up slowly a little at a time. If he starts quivering, I will stop and let him relax. After a few steps back, I let him stop and give him a treat. His head is down and relaxed, and he is standing straight.

I ask him to back again by lightly bumping his nose with pressure on the rope and he tries hard to do what I am asking by stepping back, but he is nervous and tiptoes with his hind hooves. I tell him to relax and reward him for those few steps back. I ask again, and he slowly steps fairly straight off the trailer with his hind legs (**Figure 10**). The owner and I laugh at him stepping down on his rear tiptoes.

I give him a little grain while he is standing half off the trailer, and when I say, "Back," he backs all the way off. He looks like he wants to jump back in the trailer so to avoid the temptation, I back him up farther outside the trailer with very light pressure on the leadrope. The string is not in his mouth anymore, but he backs up as if it is still there. His head is down and relaxed, and he is chewing his mouthful of grain. I don't want to rush him. The more time I spend with this, the better off the owner will be.

I load him again easily and make sure he stands squarely and not angled. I keep his head straight ahead and ask him to back with a loose leadrope. He takes a few steps back and when he starts quivering, I stop, let him relax, and reward him. If I were to push him only one time when he wasn't ready and he got hurt, it would ruin everything he just learned. I ask him to back out, and he tentatively steps out with his hind legs. He is calm, and I back him all the way out.

I want the owner to load and back her horse out the next time while I am with her just in case anything goes wrong. I caution her not to hold the leadrope close to his head because it telegraphs to him that she is nervous. You can't help but transmit nervousness to a horse when you lead it close to its head. To show how to get the horse to follow, I step back, raise a hand to him, and he turns to me, just like he did in the roundpen.

She loads the horse, and I help her a little because the horse is not used to her timing, pressure, or voice, and it makes a difference. When she wants to back him out, I tell her to hold the rope in her left hand, square off her shoulders facing his rear, and tell him to back

before pulling on it. She backs him to the end of the trailer and lets him relax. It's good that she is consistent and does it the same way I taught the horse. I show her how to keep his head out of her space with a raised arm, but he stands straight in the trailer now, not angled.

I remind her that she can relay tension to him if she gets too anxious. It is also possible to feel his tension coming through the leadrope when backing him up, and she acknowledges she has felt it before.

She is stroking him, and I encourage her to make this the most positive experience of the day for him: Try to make him just love standing at the end of the trailer. She holds his treat low so he will put his head down for it, and his eyes soften. He's not quivering, throwing his head around, or tense about backing out anymore. He is calm, cool, and relaxed. He doesn't hesitate to back out for her, and it's obvious she is in control. When he tiptoes off the trailer, we both laugh and enjoy the moment. The owner is ecstatic.

SUMMARY

The owner is amazed at her horse's progress. She is happy that she can back him out of a trailer by herself, which will be so much safer for him and anyone else who unloads him. She had been skeptical that anything would work with him and was surprised when he learned so quickly, backing out in less than an hour. She is more confident now and so is the horse.

I don't want to work the horse too much in his first lesson. The owner can practice backing him out again when she gets home. She should practice backing him with a loose leadrope before she loads him and treat every so often but gradually reduce treating him as he gets better. I remind her to stay positive and

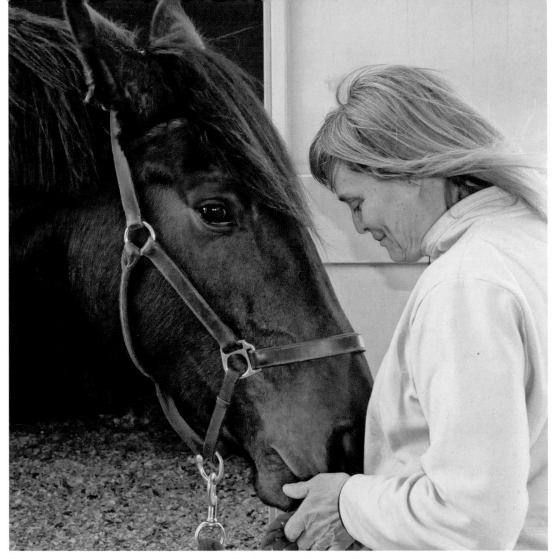

An owner enjoys a quiet, satisfying moment with her horse after a successful trailer-loading lesson.

always quit on a good note. For her to come away feeling this good about her horse when she was so frustrated with him that morning is gratifying.

> **Total time for this session: 42 minutes**

Dennis directed these exercises:

1. Taught the horse to step into the trailer on command and stop halfway in.
2. Taught the horse to step all the way in the trailer on command.
3. Taught the horse to stand aligned straight in the trailer, not throw its head, step back a couple of paces on command, and stop.
4. Taught the horse to back out of the trailer on command halfway and stop before coming all the way out.
5. Taught the horse to calmly back up all the way from the front of the trailer to the rear and back straight out of the trailer using only a verbal command.
6. Coached the owner how to back her horse out of the trailer.

INDEX

Dwain Snyder

Michelle Pettit

My horse was so dangerous, one trainer said put him down.

"Dennis Brouse changed him."

My gelding, Red, broke the hand of my trainer's assistant, and the trainer said to put him down. Then I met Dennis Brouse. His training methods helped me connect with Red in a whole new way — and changed him into a gentle, respectful horse. Dennis' methods are amazing. *Read the whole story at www.saddleupwithdennisbrouse.com*

Order your 5 disc training DVD set today. Call 1-800-285-1070 or go to www.saddleupwithdennisbrouse.com

SADDLE UP WITH DENNIS BROUSE

* Don't miss the *Saddle Up with Dennis Brouse*™ series on public TV! *

SPONSORED BY:

223

Dennis Brouse. *Nick Vedros, MindFire Communications*

ABOUT THE AUTHORS

Dennis Brouse is the host of the PBS television show *Saddle Up with Dennis Brouse*. He has been training horses since the age of fourteen, when he ran a summer riding program at a ranch. He does not "break" horses but rather focuses on cooperation and partnership. Dennis has spent thousands of hours working with horses to fine-tune his training system and has been offering training clinics since 2000. Dennis's five-disc training DVD set is available for sale from his website, www.saddleupwithdennis brouse.com. He lives in Plattsmouth, Nebraska.

Fran Lynghaug is the author of *Horses of Distinction* and *The Official Horse Breeds Standards Guide* and is the editor of *Your Horse: The Illustrated Handbook to Owning and Caring for Your Horse*. Her horse-related articles regularly appear in *Today's Horse Trader, Valley Equestrian Newspaper,* and other publications. She got her first horse at the age of eighteen and has been active in breeding, competitions, and training ever since. She lives in Downing, Wisconsin. You can find out more at her website, www .equestrian-horses.com.